Designer's Guide
to the Cypress PSoC™

Designer's Guide
to the Cypress PSoC™

by Robert Ashby

AMSTERDAM • BOSTON • HEIDELBERG • LONDON
NEW YORK • OXFORD • PARIS • SAN DIEGO
SAN FRANCISCO • SINGAPORE • SYDNEY • TOKYO

Newnes is an imprint of Elsevier

Newnes

Newnes is an imprint of Elsevier
30 Corporate Drive, Suite 400, Burlington, MA 01803, USA
Linacre House, Jordan Hill, Oxford OX2 8DP, UK

 Recognizing the importance of preserving what has been written, Elsevier prints its books on acid-free paper whenever possible.

Library of Congress Cataloging-in-Publication Data

(Application submitted)

British Library Cataloguing-in-Publication Data
A catalogue record for this book is available from the British Library.

ISBN-13: 978-0-7506-7780-6
ISBN-10: 0-7506-7780-5

For information on all Newnes publications
visit our website at www.books.elsevier.com.

05 06 07 08 09 10   10 9 8 7 6 5 4 3 2 1

Printed in the United States of America

To my lovely wife Camille

and

my wonderful parents Coy and Bonnie

Contents

Contents

Contents

Contents

Foreword

When asked if I would write the foreword for this book, I responded that it would be an honor. Robert is an engineer's engineer, and to have any association with something he does is exciting.

When it comes to programmable systems on a chip, the design world separates into three groups:

- Those that immediately get it;
- Those that eventually get it;
- Those who will never get it.

My first introduction to the Cypress Programmable System-on-Chip™ (PSoC) was in the summer of 2000, when I came by a startup company developing this new "PSoC" IC. I immediately got it! I was so impressed with the concept that I left a job as a Design Engineer to become their second PSoC Application Engineer. I must also say that Robert also got it, right from the start.

My first contact with Robert and his brother Darren was in the winter of 2002. (Logan has only three seasons; July, August, and winter.) I was on a phone conference with their company, Icon Fitness. They were looking to see if we could reduce the cost on their pulse detection products. After asking several detailed questions about pulse detectors, Darren concluded that I must have some experience with pulse detection. When queried, I explained that in my previous job I had designed several ECG front ends and was quite experienced at getting signals off a body. His response was, "You're coming out to see us, right away!"

A few days later I was on a plane to Logan, Utah to visit their company. My first impression of the company was the sign on the front door saying that no firearms were allowed in the building and were best left in the car.

From the first meeting it was obvious that Robert and Darren were both extremely capable engineers. Soft spoken, but commanding, both were open to new ideas and willing to take risks just to satisfy their curiosity. I wonder what personal hell their mother went through raising these two boys. I imagine that no motor from any home appliance was truly safe. I have always believed that any true engineer was just plain lucky to survive their teenage years. Mrs. Ashby can be proud that both her boys survived with all their fingers and most of their hearing. Their curiosity was again confirmed when, coming back from lunch, another employee was out showing off his new hybrid electric car. We immediately went over to see this obviously cool thing. He had the hood open and gave us such a look of horror upon seeing the two Ashby brothers. He asked for them to please not take it apart.

My first day with Robert was spent troubleshooting his PSoC design. He had done a remarkable job of coming up to speed on the part. We must have reconfigured the system design a couple of dozen times to test the signal path. We were successful, and a day after I went home I found out that Robert wrote an article about PSoC for chipcenter.com. Icon placed an initial order for 20,000 parts. This made Icon one of our first major customers.

Robert was one of the first believers in PSoC when believing carried professional risk. I appreciate him giving us a chance. Robert, and engineers like him, have proven that configurable analog and digital resources, coupled with an MCU on a single chip, is the future of embedding micro system design.

For those that "get it," you will enjoy this book. Robert is a good writer and an excellent teacher. For those that will eventually get it, Robert's writing should be very persuasive. If Robert cannot convince you, then nobody will. For those that will never get it, I'm sorry. New technologies have always come along and those who cannot adapt eventually stop being design engineers. It happened with the introduction of transistors, of integrated circuits, and of programmable logic devices. It will happen with programmable mixed-signal systems on a chip.

I know of no engineer outside of Cypress Semiconductor Corporation, more knowledgeable and experienced. I know of no engineer with greater integrity. This book is worth everything you paid for it. I say "you" because I expect Robert to give me a free copy.

Dave Van Ess
Principal Application Engineer
Cypress Semiconductor Corporation
Lynnwood, WA

Acknowledgments

This book is a culmination of all those who have influenced my life for the better. I owe my drive to succeed to my parents. My father always encouraged me to set my sights high, and my mother never doubted that I could reach the stars. Both my father and father-in-law are excellent teachers. I've been fortunate enough to experience the crowning event beyond making something work, which is being able to watch someone you've taught make something work.

This book was completed in part as a tribute to all the generous authors who have written application notes and methods in varied strains of life. I would not have the knowledge that I do if such charitable persons were not willing to take their own time and effort to provide their knowledge for the benefit of others with no immediate reward for their efforts. I thank these unsung heroes.

There are many people in the professional field who have helped me improve on my talents to make this book possible. I would like to thank those superb engineers at Cypress Semiconductor Corporation for designing such a fabulous series of parts. There are many people there who have left an indelible impression on me that helped to give me the confidence and resolve to see this book through. Among those PSoC persons I rate as extraordinaire are Ken Dees, for his encouragement, endless knowledge and help with having Cypress support backing my book; Dave Van Ess for his patient explanations and native joviality; Mohana Koteeswaran for her continued support on my projects; Jon Pearson and Matt Basinger for providing permissions to much needed material; and George Saul, living proof that good things can come from Utah. I also would like to thank Carol Lewis and Tiffany Gasbarrini for their belief in my work and opportunity for writing this book.

Finally, I would like to thank my wife and family for the support and encouragement to help me to the finish line. Since writing is far from my favorite tasks in life, I would never have completed this book without their help.

What's New with PSoC?

The popularity of the PSoC processor is growing more quickly than ever before. Application notes are being added all the time, new parts are being introduced, and the tools are more powerful.

I would like to give a brief overview of a couple of parts that Cypress has since introduced that are sure to be huge successes on future designs. The first part is the CY8C24273. This part integrates a full-speed (12 Mb/s) USB peripheral functionality along with the PSoC core. Cypress began earlier by making the enCoRe™ USB controller. They were part of the CY7C637xx series and they gave you RAM and ROM to run the controller and manage some simple I/O. They were selectable from within PSoC Designer as a part that you could configure and place modules and their associated APIs; very similar as to how you would configure a normal PSoC part. However, the PSoC configurable digital and analog blocks weren't there. There were only dedicated blocks that could be one thing and one thing only. This allowed you to have the USB communication along with a PS/2 controller and SPI port. However, the ease of use for a USB device was terrific. Cypress even found a way to tune its clock from the USB and eliminated the need for an external crystal.

The CY8C24273 combines the easy to use USB interface with a normal PSoC core. There are four digital blocks and two columns (six blocks) of analog. It has 16K of Flash and 1K of RAM. (The USB function works from a dedicated RAM buffer.) The PSoC core still runs at 24 MHz. When used in a USB system, the PSoC can fine-tune the internal oscillator to achieve a much higher accuracy for the internal main oscillator. Cypress has added another MAC into the equation giving you two

8 × 8 multipliers with 32-bit accumulators. The rest of the resources within the part utilizes the improvements of parts in the past, including the improved analog capabilities and the chip-wide analog mux system, which allows you to put analog signals on every pin. The make-break circuitry is available to even enable your system with the capacitive touch sensing of the 21xxx parts.

The second part that has been announced in the Cypress lineup is the wirelessUSB™. Cypress has made an inexpensive, but powerful radio system that allows you to cut the cord. The first radio parts operate via a communications port and can be tacked onto your PSoC system to give you wireless capabilities. However, the CYWUSB6953 part that is coming will integrate the wireless radio right into the PSoC part. This new strain of processors will be called the PRoC™ (Programmable Radio-on-a-Chip) series. The CYWUSB6953 is targeting low-cost applications so the PSoC core section of the chip has four digital blocks and two 'Type E' limited analog columns as is found in the 21xxx parts. The radio capabilities of these parts are really exciting. The radio operates in the 2.4 GHz ISM (industrial, scientific, medical) band. It operates on its own protocol so it won't interfere with other radio devices in that range. In fact,

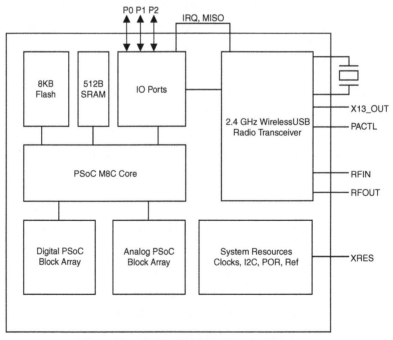

Figure 1: CYWUSB6953 Block Diagram

testing has shown that it is much more robust than many devices that operate in the same range.

All that is needed with your part to make it operational is a crystal, a few capacitors and a couple of inductors. The footprint is very small. The antenna can be fashioned with traces on the PCB leaving you with a small, tidy package that is ready to cut the cord and allow you into a whole new world or product possibilities.

Cypress is on the move. Their combination of functionality, ease, and low cost assures them a leading position on future microcontroller design. The parts mentioned previously are moving to full production quickly and should be available by the time you read this. If not, contact your local Cypress representative for more information. As always, updated information can be found at the Cypress website: http://www.cypress.com.

What's on the CD-ROM?

The accompanying CD-ROM is divided into four categories:

Application Software

- Contains the latest PSoC application software available at the time of this writing. Check http://www.cypress.com for the latest versions.

- Order of installation:

  1. Install PSoC Designer

  2. Install the Service Pack

  3. Install PSoC Programmer

  4. Install PSoC Express

Application Notes

- Contains a collection of application notes from the Cypress website.

Design Catalog

- Contains a group of design templates that were available with older installations of PSoC Designer.

Data Sheets

- Contains a collection of data sheets for PSoC information.

Introduction to Microcontroller Basics

The purpose of this chapter is to explain the basic structure and inner workings of a microcontroller. It is intended for those readers who may not have much experience with microcontrollers or who are getting their first taste of microcontrollers with the Cypress Semiconductor Corporation Programmable System-on-Chip™ (PSoC™). I will forgo the somewhat traditional explanation of the history of computers as that information can easily be found by a quick search on the Internet. The important parts that I would like you to retain from the history are as follows:

- Semiconductors are made from silicon. They can be conductive or nonconductive, which allows you to form a basic electrical switch.

- Semiconductor circuits can now be made extremely tiny allowing us to put millions of such switches on a single chip.

- Most computers today work with a combination of ones and zeros to accomplish their task. Every piece of information used in a computer is stored as ones and zeros.

What is a Microcontroller?

A microcontroller is a microprocessor system with attached peripherals. Think of the CPU in your computer—it's a microprocessor system. I like to use a computer system to explain microcontrollers because it is a very similar system that is familiar somewhat even to many grandparents in today's world.

Your computer most likely has an Intel or AMD chip. You might be one of the Apple adherents, which would mean that you would have a Motorola chipset. The CPU in your computer does do the thinking, but in order to make that CPU useful, it needs some other parts linked to it. Following is a list of some of those parts, along with their function:

CPU – This is the brains of the system. The central processing unit (CPU) knows how to read and write information to the various areas of memory. It also can perform logical instructions on numbers. The most basic and universal are the abilities to add, subtract, logical OR, logical AND, logical exclusive OR (XOR), shift, move, and copy numbers. Some processors will perform a few more complex operations than these, but they are all derived from these simple actions.

Cache Memory – The cache memory is closest to the CPU in terms of location and speed. It is sometimes integrated right on the same chip. Note that this doesn't necessarily mean that they are the same piece of silicon on the inside, but are simply housed in the same package.

RAM – The RAM is next in line in terms of speed. The term RAM is an acronym for *random access memory*. The acronym has lost some its individuality as most memory today can be accessed in any order.

Hard Drive – The hard drive is last in the line of speed within your computer system. It is the area where all of your programs are stored and are not erased when you power down your computer. It is also typically the largest storage area.

The relationship of these components can be likened to a student. Imagine that you are sitting at a desk and you are a computer. Your task is to complete a math assignment in front of you that is quite complex. You have a calculator on the desk. It is your CPU. The piece of paper immediately under your hand is the cache memory. All other pieces of scratch paper around your desk or any book of reference open on your desk is the RAM area. All information (books, previous assignments, and so forth) that is on the bookshelf next to the desk represents the hard drive.

If I know what numbers I need to use with the calculator, they will be on the paper under your hand, and you can enter those numbers in the calculator and write your results on that paper. If you want to use a different piece of paper on your desk, you need to first fetch the paper and position it under your hand. If you need to work from something that resides on the bookshelf, you will need to get up from the desk to retrieve that information, place it on the desk while you are working on that information and then (assuming you are tidy) return that information back to the bookshelf.

This picture is very similar to the computer. Retrieving information from the hard drive takes much longer than retrieving information from cache memory or even from RAM. You must bring the information into the RAM or cache memory in order to manipulate it according to your program. You are then able to take that memory and store it back out to the hard drive.

The reason that I draw this parallel with computers is that there are also very similar parts on a microcontroller. Each of these parts can be taken for granted and forgotten. If you don't learn the structure of the microcontroller that you are dealing with, you can not be the most efficient with its usage. You also may be prone to make some errors with your programming that become very difficult to solve without knowledge of the microcontroller's structure. Let's look at the PSoC equivalents to our computer system.

> CPU – M8C CPU core
>
> Cache Memory – CPU registers A and X
>
> RAM – RAM
>
> Hard Drive – Flash Memory (ROM)

Each of the PSoC components has the same basic function and capabilities as the computer system described with our math assignment above. The line between the cache memory and RAM is a little fuzzier, as many microcontrollers today will allow you to perform math functions on RAM directly without having to move that memory into the CPU registers first. An x86-based computer system doesn't work that way, but the PSoC sports some of that capability.

What About Peripherals?

Peripherals are what distinguish a microcontroller from a microprocessor system. If we were to go back and compare the computer system, such devices are typically found outside the chip. You might compare a parallel port on a computer to the I/O ports of a microcontroller. That parallel port is actually a separate chip that communicates with your CPU within the computer. The PCI slots likewise are handled with separate chips that communicate with the CPU. The same scenario exists with USB ports, FireWire ports, interrupt controllers, and so on.

The PSoC microcontroller contains everything needed to run an entire system. The Flash, RAM, CPU, ports, and configurable blocks communicate with the CPU as separate systems. However, these subsystems and their communication paths are all conveniently contained in one chip. The configurable logic allows you to create the serial ports, timers, PWM generators, and other devices without adding additional circuitry.

The technical reference manual contains a block diagram that illustrates this structure.

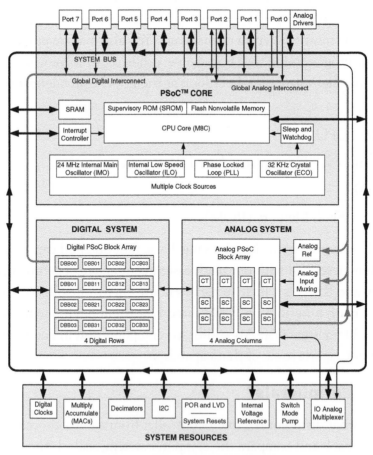

Figure 1: PSoC Block Diagram

What's in the CPU?

The CPU is the brains of the system. It contains several subsystems. The important subsystems that I want to point out are the program counter, the instruction decoder and the ALU section. The program counter is a subsystem that looks at a specific address within the Flash memory to return instructions and data. Specific logic will use the program counter to decide what bytes in Flash to feed into the instruction decoder. The instruction decoder contains logic that decodes the numbers returned from Flash to decide what instructions you want to perform. These instructions tell the CPU what to do next. Remember that the computer can add, subtract, AND, OR, XOR or shift numbers. It can also change the location of execution. If the instruction is to change the location of execution, then the program counter is loaded with a new address and execution will begin with that new address in Flash. If the instruction tells the CPU to perform some logic then the appropriate numbers are sent to the ALU (arithmetic logic unit) to be manipulated accordingly.

The microcontroller also contains the peripherals that are seen outside the CPU in Figure 1. The CPU area has the ability to control these peripherals according to the instructions that it receives.

What Can a Microcontroller Do?

The microcontroller combines its CPU capabilities with the peripheral control and embedded memory to create a complete system. It will respond to specific inputs coming from I/O pins, timers, communications, and so forth, and manipulate the information to generate the appropriate output signals or information. The I/O pins give the microcontroller the ability to read buttons and logic states from other devices. The I/O pins also can output signals to turn on lights, run motors, and drive display devices. The timers, communication modules and analog-to-digital converters allow the microcontroller to perform special tasks such as communicate with other systems like your desktop computer and read special signals like a device that may read temperature. The Cypress PSoC is called a *Programmable System-on-Chip* because it contains adequate resources to require very little external circuitry.

Most microcontroller designs are small 8-bit processors that are found in toys, appliances, and a variety of smaller electronics. There is an unlimited number of possibilities of how a small 8-bit microcontroller can be put to work. Since the cost of these microcontrollers keeps dropping so dramatically, the microcontroller is always finding its way into new applications to make life more comfortable, more efficient and just plain-old more fun!

How Does a Microcontroller Work?

A microcontroller is created by combining thousands or millions of electrical switching devices. Just as the programmer must simplify their complex operation into the AND, OR, XOR and shift instructions to accomplish the task, the microcontroller designer must decide how to create a device that can accomplish those simple tasks using a series of electrical devices, for example, transistors, FETs, and diodes. Most microcontrollers operate on a binary system. In a binary system, there are two defined states. These states are referred to in various ways. Among the more common titles are one and zero, high and low, or on and off. In order to understand how these devices can create a microcontroller, you need to know some basics about numbering systems and logic.

A Little Bit About Numbers

Microcontrollers handle data in a digital format. It's a bunch of ones and zeros. In order to understand the logic of microcontrollers and to work with them effectively, you will need to be conversant in at least two other numbering systems besides the decimal system that you worked with all your life. These number systems are binary and hexadecimal. The binary and hexadecimal number systems use a different base or radix. The binary system is base 2, and the hexadecimal system is base 16. (Note: the decimal system is base 10.) In order to understand how the number systems of a different base work, let's look at the decimal system that you use today.

Table 1: Base 10 (Decimal) Example

00	10	20	...	90	100	110
01	11	21	...	91	101	111
02	12	22	...	92	102	112
03	13	23	...	93	103	113
04	14	24	...	94	104	114
05	15	25	...	95	105	115
06	16	26	...	96	106	116
07	17	27	...	97	107	117
08	18	28	...	98	108	118
09	19	29	...	99	109	119

Table 1 shows how to count in the decimal system. I haven't included the numbers from 30–89 as I'm sure that you are able to fill the in blanks. The first column might look a little funny as you don't usually print a leading zero. I've done this to illustrate the digit in the tens place is zero for that first column. As you look at the table, you notice a pattern. Each column has one value for the tens digit as the digit in the ones place goes from 0 to 9. Once I run out of values to put in the tens digit place, then my hundreds digit will change from 0 to 1, and I will start the pattern over again. Each time you run out of combinations with the available amount of digits, you simply add a digit. If you can understand that basic table, then you can understand the numbering system of every base.

The decimal number system is base 10. That is why I have arranged the table to have 10 rows. This table will simply change the number of rows as I change from one base to another.

Table 2: Base 2 (Binary) Example

0	10	100	110	1000	1010	1100	1110
1	11	101	111	1001	1011	1101	1111

Table 2 depicts the base 2 numbering system. The base 2 system is also called a *binary system*. The base 2 system is shown here in a table having only two rows, as there is only a 1 or a 0 that can occupy each digit space. Table 2 shows 16 different numbers. The number elements in the binary system are referred to as bits more often than digits. Therefore, you can see that it takes 4 bits to show the 16 different numbers in Table 2.

Table 3: Base 16 (Hexadecimal) Example

0	10	20	30
1	11	21	31
2	12	22	32
3	13	23	33
4	14	24	34
5	15	25	35
6	16	26	36
7	17	27	37
8	18	28	38
9	19	29	39
A	1A	2A	3A
B	1B	2B	3B
C	1C	2C	3C
D	1D	2D	3D
E	1E	2E	3E
F	1F	2F	3F

Table 3 shows a base 16 or hexadecimal numbering system. Letters are used in place of numbers to illustrate the additional 16 values of each digit. Hexadecimal is often used in showing values used in microcontroller programming as it gives a numbering system that works well with microcontroller architecture, but allows you to look at a consolidated notation without getting lost in a long string of ones and zeros.

It is an important skill to understand the different numbering systems and work enough with the numbers to be able to make simple conversions back and forth between these different systems. Understanding how the numbering systems work and their individual characteristics will greatly assist you in writing effective code and algorithms. Efficiency with these numbering systems comes with lots of practice and lots of playing with numbers. A good spreadsheet program, or using the scientific view of the calculator that is included with the Windows® operating system gives you some immediate conversion tools. There are several times when lessons learned from number games or neat math tricks find their way into my own work. Look for these patterns and tricks. Also realize that many patterns and tricks will apply across different numbering systems when the base is taken into account.

Table 4: Number Conversion Table

Decimal	Binary	Hexadecimal	Decimal	Binary	Hexadecimal
0	0	0	25	11001	19
1	1	1	26	11010	1A
2	10	2	27	11011	1B
3	11	3	28	11100	1C
4	100	4	29	11101	1D
5	101	5	30	11110	1E
6	110	6	31	11111	1F
7	111	7	32	100000	20
8	1000	8	33	100001	21
9	1001	9	34	100010	22
10	1010	A	35	100011	23
11	1011	B	36	100100	24
12	1100	C	37	100101	25
13	1101	D	38	100110	26
14	1110	E	39	100111	27
15	1111	F	40	101000	28
16	10000	10	41	101001	29
17	10001	11	42	101010	2A
18	10010	12	43	101011	2B
19	10011	13	44	101100	2C
20	10100	14	45	101101	2D
21	10101	15	46	101110	2E
22	10110	16	47	101111	2F
23	10111	17	48	110000	30
24	11000	18	49	110001	31

Basic Logic

The basic building block of microcontrollers and most computers today is an electronic switch. Millions of these switches are organized to allow a microcontroller to work properly. The basic switch has two states: ON and OFF. A series of switches are designed to allow the microcontroller to perform basic logic instructions.

Using an ordinary light switch, if the switch is closed, the voltage is transferred and the light goes on. (Note: For the purposes of this discussion, forget about 3-way and 4-way switches that can be found in your home.) This basic switch can be used to show how logic works. In electronics, there are many items that can be thought of as a switch just as you see with the example of the light switch. Depending on the design of the switch in electronics, the ON state may require a high signal, and the OFF state requires a low signal, or it may be reversed. In the reversed switch, the ON state requires a low signal, and the OFF state requires a high signal. In order to compare this to the light switch example in your home, just think of the electrician mounting the light switch upside down.

We can use a combination of light switches to illustrate the basic logic that is used in a microcontroller. We will look at some basic logic using light switches. There will be two inputs for one logical output result. Binary numbers allow four different possible combinations with two elements. The two elements will be labeled 'A' and 'B' and are represented by the light switches. The output is depicted by the light bulb. Flipping the switch up is considered a '1' or logic high, and flipping the switch down is considered a '0' or logic low. The light bulb output is considered a '1' or logic high when it is on. The light bulb output is considered a '0' or logic low when it is off. Consider the following illustrations.

OR Function

The OR function will output a high signal if input 'A' or input 'B' is high. Likewise, if both input 'A' and input 'B' are high, a high signal appears at the output. If neither 'A' or 'B' is high, then a low signal is seen at the output.

OR Function		
A	B	Output
0	0	0
0	1	1
1	0	1
1	1	1

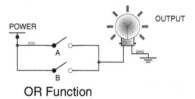

OR Function

AND Function

The AND function requires that both 'A' and 'B' are in a logic high state before the output state will be high. If only one input is high, or if neither input is high, then the output state will be low. This is illustrated by the truth table for the AND function.

AND Function		
A	B	Output
0	0	0
0	1	0
1	0	0
1	1	1

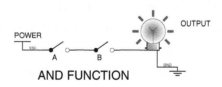

AND FUNCTION

NOT Function

The NOT function is simply a way to invert the input signal. It can be used in conjunction with the other logic to provide more complex logic. It only has one input.

NOT Function	
A	Output
0	1
1	0

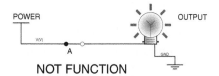

NOT FUNCTION

Combinations of these basic building blocks allow the designers of microcontrollers to build more complex blocks of logic. These blocks of logic are used to create adders and multipliers. They also create latches, input pins, and output pins. If the simple switch is the cell of a microcontroller, then the logic blocks are the molecules.

Instructions and Data Handling

The instruction decoder of a microcontroller allows for a predefined set of instructions. In the PSoC microcontroller, there are 37 instructions. All programming must eventually be broken back down to a series of these 37 instructions. If you are programming the PSoC in C, then the C complier will convert your C language back down into these 37 instructions. Among the 37 instructions are the basic logic instructions that we have already discussed, such as AND, OR, XOR and shift. There are also some more complex instructions. There are some logic modules contained with the CPU that will complete these more complex instructions without having to break them down into more simple logic. Some examples of these more complex instructions are instructions that add, subtract, multiply, and compare.

Instructions are stored in memory as coded values. The programmer doesn't have to remember the coded values to describe the instruction. The assembler will accept a mnemonic, which represents the instruction. A mnemonic is a short abbreviation or word which describes the instruction. These instructions can be followed by one or two operands. The operands can describe what is to be added, shifted or where to move program execution.

The first operand can be thought of as a destination. If the instruction is making a change to the location of program execution, then the operand contains information on where the new wanted address is located. In a complement operand instruction, no other information is needed to describe the instruction, so only one operand is needed. If there are two operands, then the second operand can be referred to as the source. Consider the following instruction:

```
mov [2],35
```

In this instruction, the mnemonic given is a move command. This command will move the number 35 into the memory location at address 2. The second operand of the instruction is considered the source because it is the number or location of what is being moved. The final location or destination of that instruction is in the first operand. The first operand in this instruction is an address.

Operands within the instruction are given as numbers or labels that represent numbers. When you list an operand or assign a value to a label, the default base for the number given is decimal or base 10. If you desire to use a different numbering system, then you will need to use the appropriate assembler notation to describe that number. For example, if my instruction of mov [2],35 was changed to mov [2],35h, then the register at location 2 would have an entirely different value in it. The first instruction loads a 35 into location 2. The second instruction has an 'h' appended to the number. This signifies to the compiler that you are using hexadecimal notation. Therefore, the value of 53 (decimal equivalent of 35 in hex) is loaded into the register.

A complete list of all the instructions available in the PSoC is found in the help file of PSoC Designer™. This help file is opened by opening **Help Topics** from the **Help** menu within PSoC Designer. Inside the **Contents** tab, open the **Assembly Language Reference Book** and select the book called **Instruction Set**. This will show you all the commands that the CPU recognizes. If you program in C, or use other macros, a list of these commands will be combined to perform the potentially more complex action listed in your C code. Following is an example:

```
C Language    Assembly Language and comments
F = D + C    mov    A,[D]    ;move D to accumulator where I can add
             add    A,[C]    ;add C to D, result is left in accumulator
             mov    [F],A    ;move result to F
```

Addressing Methods

There are various methods in microcontrollers to address memory locations. The methods are referred to as addressing modes in the PSoC. The help file for the PSoC mentions ten different addressing modes, but there are only four different concepts to learn. The ten different modes show these four concepts in different positions and combinations. The ways to use these methods are not universal for all microcontrollers, but this does depict how these methods are used with the PSoC. These concepts are listed here with a short description:

Immediate – The immediate instruction contains a value to be used in the instruction as one of the operands. The immediate value is given in the operand with no other punctuation or symbols around it. It can also be an expression. Since the immediate is a simply a number and doesn't represent a RAM location or register, it can only be a source value. An example of an instruction using an immediate is the mov [2],35 listed earlier. The value of 35 is the immediate value. An expression can also be used for the immediate value. For example, mov [2],35*2 will place the value of 70 in the RAM location 2.

Direct – Direct addressing refers to instructions where the source or destination is contained within square brackets. The square brackets tell the compiler that the destination is a RAM location. If the square brackets are preceded by the abbreviation reg, then the compiler knows that you are addressing a register location separate from the general-purpose RAM, as seen in the examples below.

```
mov    [2],35      ;move 35 into RAM location 2
mov    reg[2],35   ;move 35 into the register location 2
```

Registers are used in the PSoC for storing values that configure the user modules or the CPU. It is in reality RAM also, but the use of the reg abbreviation allows the separation of addressing this different area of memory without other complexities.

Indexed – Indexed addressing refers to instances where the number or expression is given points to the final destination or source information. The X register is used in the PSoC for indexing to other locations. When the X register is used without the square brackets, its use is similar to the accumulator. However, when you use the X register within square brackets, then you switch to the indexing method.

```
mov    X,35          ;move a 35 into register X
mov    [X],15         ;move a 15 into location 35
```

In the example above, the second instruction will use the current value contained within X as the location of the final destination. Since I had moved a 35 into the X register before using the indexing method, 35 became the final destination location.

PSoC also allows you to have an offset for the indexing method. This can be useful when you are loading tables. Let's say that you have a table of values that is three values wide as depicted here:

10	11	12
20	21	22
30	31	32
40	41	42

You would have to store this table sequentially in RAM, that is, every row would begin at an address that was three higher than the previous row. If I wanted to bring back the values of a particular row, then all I need to do is to get the X register loaded with the address of the first element in the desired row.

```
mov    X,FirstRow    ;load X with the desired row address
mov    A,[X+0]        ;A gets the first element
mov    [FIRST],A      ;store the first element
mov    A,[X+1]        ;A gets the middle element
mov    [MID],A        ;store the middle element
mov    A,[X+2]        ;A gets the last element
mov    [LAST],A       ;store the last element
```

This example makes the use of indexing seem superfluous, but as the table grows in size and complexity, the indexing method can be extremely useful.

Indirect Post Increment – There is only one instruction that utilizes the indirect post increment addressing method. That instruction is mvi. The mvi instruction has the accumulator as the source or destination. A direct address RAM location is the other operand of the instruction. The contents of the RAM location points to the memory location used as the source or destination for the instruction and the RAM location is then incremented after the move has occurred. Following is an example:

```
mov   [2],35   ;move a value of 35 into RAM location 2
mov   A,10     ;load a value of 10 in the accumulator
mvi   [2],A    ;Copy the 10 to location 35, location 2 goes to 36
mov   A,20     ;load a value of 20 in the accumulator
mvi   [2],A    ;Copy the 20 to location 36, location 2 goes to 37
```

A Little Bit About Paging

The newer PSoC parts provide the user with more RAM than 256 bytes. When an 8-bit processor grows in RAM size larger than 256, the designers are faced with three choices: change the core to allow a larger address structure, increase addresses to be more than 1 byte, or implement paging.

The redesign of the core moves you into an entirely different part, which may not make sense if you simply want to allow the user to have more RAM. You can increase the address size of all RAM to be multiple bytes, but now you bloat your code usage as you are using another byte of ROM for every instruction that addresses a RAM location. If a RAM location is used as both the source and the destination in the instruction, then you are adding 2 bytes to these instructions.

Paging is having a designated register point to which block of 256 bytes you want the instructions to be addressing. You can think of this register as the added addressing bits needed to specify the increased RAM space. The advantages of paging are that it requires minimal impact to the compiler, microcontroller core and it doesn't bloat the code size. However, paging can become a large pain if you have to switch pages frequently. Poor design can lead paging to actually increase the code size more than adding another byte of address in some situations.

PSoC decided to use the paging approach in their larger micros. The key to making the paging method successful is to keep in mind that you want to switch the pages of RAM as few times as possible. This will minimize the amount of added code due to page switching and will also minimize the common error of altering RAM on the wrong page. That can present some very difficult bugs to find.

More Information

Microcontrollers are best learned by action. Don't be afraid to get your hands dirty, so to speak, and dive in to try to figure it out. Don't be afraid to grab a pen and paper and walk through your logic rather than sitting for half an hour with a furrowed brow trying to work it all out in your head. There are still many times when I search on the Internet, looking at application notes or ideas on how to accomplish a task. All of the basics covered here are reiterated in many places on the Internet. Go looking for them and learn them.

Don't be discouraged if you don't remember all the basics from your schooling, or if you have never learned them to begin with. Every professional that I have seen in the field cannot remember everything they learned in school. We have often had to resort to textbooks or the Internet to refresh our minds on fundamental algorithms or basic electronics to accomplish our tasks. The truly successful designer isn't necessarily the one who knows all the information, but is the one who knows where to find it.

Why Use the Cypress PSoC?

Electronics have dramatically altered the world as we know it. One has simply to compare the conveniences and capabilities of today's world with those of the late 1900s. Computers have gone from the size of a room to the size of a pocket planner. Gone are the days of seeing a telephone only in the cars of the rich. Today, you see people in all walks of life with multiple cell phones. The advent of the Internet has brought the entire world a few steps closer and moved the target for tomorrow's electronic miracles out to seemingly sci-fi levels.

Much of what is possible with electronics today is due in part to integration of large circuits into smaller and smaller packages. Millions of transistors are etched into a fraction of a square inch of area and a paper thin thickness. Two advantages of this integration have been to allow electronics to be shrunk such that portable complex devices such as PDAs and cell phones can exist without having to lug around a 6-pound box full of bulky components. The other major advantage is that it dramatically cuts the expense of electronic production. The silicon wafer and packaging expense comprise much of the significant costs of building microcontrollers today.

As the cost of microcontrollers continues to drop, you will see that more powerful brains in existing designs will result in increased functionality and more flexible operation. There will also be a huge increase of new designs that have never before existed, since there are dozens of micros I've been shown that can be bought for well under a dollar in quantity. These micros boast such features as analog-to-digital conversion, Flash memory that can be self programmed, multiple timers, built-in hardware-based communications, and LCD drivers. Such capabilities have been driving research into exploring such products as tattoos that can be altered to display what you want, clothes with brains built into them to change their look or insulation characteristics, and smart implants to dispense medicine, restore lost body functions to paralyzed victims, or build the basis to give sight to those blind from birth.

Cypress Semiconductor Corporation was formed to introduce the Cypress PSoC™ family. The product has been out for a few years now and is rapidly gaining popularity. The Cypress PSoC family offers a unique blend of adaptability in design for a relatively low price. In today's world of ever increasing automation and digitization of the world around us, the PSoC family allows you to add brains and capable signal conditioning to a design in one complete chip. This level of integration previously existed only in much larger parts, making the advantages of integration pale in comparison to the added price and complexity. Past designs with analog needs would amplify, filter and condition the analog signals using dedicated analog circuitry and then use the micro's analog-to-digital converter to read in the signal and do any final digital manipulations in the micro itself.

Digital manipulations of an analog signal can be quite processor-intensive and limiting to smaller micros. The Cypress PSoC adds analog manipulation capabilities into the micro for a complete one-chip analog/digital design. The PSoC chipset is designed with programmable digital blocks and programmable analog blocks that contain continuous analog and switched cap analog blocks. Since the analog blocks and all interconnects are all contained inside the chip, this provides a small design form factor. The PSoC allows real-time reconfiguration of the analog and digital blocks that allow the design to perform more than one function or to adapt to perform better.

When I was first introduced to the PSoC family, they weren't yet in full production. I was intrigued by the advertised capabilities of the part and thought of several immediate projects where such a micro would be useful. Since then, I have worked on more than a dozen designs using the PSoC family. The adaptability and flexibility of the PSoC design assures it a place in current and future designs for quite some time to come.

Notable Qualities of the PSoC Family

The PSoC family offers full 48-MHz CPU operation. The internal oscillator is rated at ±2.5% accuracy. If you aren't concerned about creating a real-time clock or performing asynchronous communications, then the internal oscillator might very well suffice and you can save the price of a crystal. You can implement an external crystal if you do need the added accuracy in your design. The processor can also be run at lower speeds to allow a power savings if desired.

The 8-bit CPU is built around a Harvard architecture and has a built-in multiplier/accumulator (MAC) allowing for speedy instruction execution and single instruction multiplication.

The PSoC can be programmed in-circuit via a serial connection and is able to self-program its own Flash memory. This provides the ability to emulate EEPROM within the Flash and upgrade your system while it's in the field.

The PSoC has a variety of parts that range from six to more than sixty I/O, 2K–64K of Flash, and multiple combinations of digital and analog blocks.

My Experience with the PSoC Family

I've been able to work on many designs with the Cypress PSoC family from conception to production. These designs have various functions including signal processing, user interface, motor control, and analog measurements. These designs have given me an opportunity to try out the capabilities of the PSoC in different situations.

I have also had the opportunity to teach the Cypress PSoC at Utah State University. It has been a marvelous teaching tool for students not only to learn the basics of microcontroller design, but also to work with the flexibility of reconfiguration and the growing pains of working with an analog system. Many of my students have abandoned their previous choices for micros in favor of the PSoC due to its ease of implementation, its great development tools, and powerful capabilities.

Getting Over Those Speed Bumps

The PSoC family of processors was breaking into some new territory when their parts were first designed. I found that I had some unique learning curves as I was working on making this microcontroller do things that I had not as yet tried to accomplish in any other device. The growing pains were, at the time, somewhat difficult, but the successes far outweighed the disappointments for me.

I have heard some criticism about the limitations in the PSoC family. I believe that too many can get caught up pointing out the limitations of the parts and have lost sight of the myriad of problems that this family of micros can solve. I will discuss some of the limitations that I found in the PSoC along with methods and work-arounds to smooth out your speed bumps and get your design up and running as quickly as possible. The limitations of the PSoC shouldn't discourage you from considering it for your next project.

My first designs utilized the 25xxx/26xxx series as these were the only parts available at the time. There has since been a release of a new flavor of PSoC processors (24xxx/27xxx) during the development of this book that eliminate many limitations of the previous family and adds additional features that enhance the capabilities of the previous parts. The 25xxx/26xxx parts are still available and are still the best choice for some designs that I'm involved with. However, the newer PSoC parts are filling the gaps quickly and Cypress intends to phase out the older parts with their idiosyncrasies.

The scope of this book is to give the reader a reference to use in concert with the data sheet and user manual. I intend to show methods of design that will help you to utilize the PSoC to its greatest potential with a minimum of heartaches and late nights. I will include specific examples that will highlight the PSoC's unique capabilities and have put each of those examples on the companion CD-ROM.

A True System on a Chip

Since the advent of the transistor and later the first integrated circuit (IC) that stored many transistors onto a single chip, there has been a continual drive to see how much we can fit into a single piece of silicon. The densities have been increasing and the processes have been shrinking away almost to nothing. The days of saving your pennies to get that math coprocessor for your 386 computer at home are long forgotten. Probably a good half of the technical population doesn't remember actually using an 8088 or even a 286. Moore's law has long outstripped those days of yore.

Microcontrollers aim for integration. The goal isn't, for example, to have a system to which you can add more RAM. The object of the game is to have one chip serve as the CPU, RAM, ROM, and I/O necessary to get the entire project done. Over the years, designers have been bringing more and more onto the microcontroller to make it more powerful. They started to add such things as comparators, analog-to-digital converters, timers, interrupt controllers, and even multipliers/accumulators. In the last few years, I've even started to see some other micros build in amplifiers.

The PSoC development team wanted to take the integrated microcontroller system a step further. Even though there are a myriad of projects that you can accomplish with the microcontrollers of today, there was a large group of projects that required some analog manipulation of signal external to the microcontroller to process the signals to a point that they could be handled correctly by the microcontroller. The PSoC design-ers strived to design a microcontroller that would allow you to perform the necessary analog functions within the microcontroller itself and eliminate the need for the external analog components. At the same time, they wanted to keep all of the integra-tion of other microcontroller systems. Thus, they created a true system on a chip.

The digital and analog capabilities are controlled via configurable blocks. The blocks are divided into sections according to design. Some blocks are designed to perform analog functions. Other blocks are used for digital purposes. These blocks are con-figured via a series of registers. These registers are initialized on power up using the PSoC's built-in Flash memory. There isn't a need for a separate configuration memory.

Embedding this analog functionality within the microcontroller poses several advan-tages. There is of course the space advantage. Less real estate means a smaller PCB and pennies in your pocket. The integration typically means a simpler design since you don't have to route power and ground to multiple ICs across the board. There is also the advantage of not having to route the signal lines from one chip to another, since the routing is handled within the microcontroller.

There is advantage in the noise realm. The electrical signal paths within single ICs are much shorter than they are going to be if you have to route them externally between components. This presents smaller antennas. Smaller antennas mean you gain in both noise immunity and noise emissions. Everyone I know who has spent time in the mystical realm of trying to understand the voodoo characteristics of noise appreciates the opportunity to avoid a noise problem.

Previous designs that I've worked on required additional external circuitry to power down the analog section of the circuit in order to conserve power during periods of inactivity. The PSoC micro has the ability to power down the analog section by simply writing to a register. This provides a simple way to minimize power in a mixed digital and analog design. In addition, Cypress allows you to choose different power settings within its analog systems which allow you to easily optimize your design for maximum battery life and still maintain acceptable performance.

Since the configurability of the PSoC micro is all controlled by a series of registers, the PSoC is able to change its configuration on the fly. Cypress helps to facilitate this changeover by allowing the user to specify multiple configurations in the Device Editor of PSoC Designer™. If a small change is needed, then the user is allowed to modify these registers directly, which can be performed quickly and without an extensive use of program space.

A Work in Progress

There have been great ideas for improvements for the PSoC over the last couple of years, and happily some of them have been implemented. As Cypress works to resolve existing glitches and improve subsystems of the PSoC family, the PSoC will find its way into more existing designs and will open the way for new designs in areas that have never seen the benefits of a microcontroller.

Each day as I work with the various micros in my line of work, I always know that I'll be in for a day of excitement as I work with the PSoC. I look forward to the PSoC designs that I will do in the future.

CHAPTER 2

Structure of the PSoC

M8C Core

At the heart of the PSoC is an 8-bit Harvard architecture processor that can clock out 4 million instructions per second (MIPS). The structure is straightforward, which allows a novice to have a project up and running rather quickly. It has an accumulator, although many operations can be performed directly with the registers, allowing compact code. There is one index register (X) to allow for indirect addressing. The limitation of having only one index register is offset by the ability of the M8C core to move data directly from one RAM location to another without having to use the index register or the accumulator.

The architecture allows up to 64 Kbytes of Flash memory. This Flash memory is in-circuit programmable. This means that a properly designed circuit will allow the PSoC to be programmed after it has been placed on the board. The Flash requires only normal operating voltage for programming. This Flash memory is also in-application programmable (IAP), which means that the micro is able to program its own Flash memory. The Flash memory has an erase write endurance of 50K cycles. This high endurance, coupled with a small block size of 64 bytes, allows an effective use of the PSoC Flash for nonvolatile memory in many applications.

The included 8 × 8 multiplier/accumulator (MAC) helps processor-intensive calculations such as those used in digital signal processing (DSP). It is a signed multiplier that has a 32-bit accumulator.

The processor is linked to several digital and analog blocks. These blocks provide the flexibility that separates the PSoC family from other micros. The digital blocks perform digital functions such as timers, counters, UARTs, and PWM generators. The analog blocks are used for analog functions such as programmable gain amplifiers, comparators, and filters. There is a series of interconnects that link blocks with other blocks, various clock taps and with the I/O. Successfully configuring the blocks and interconnects is the largest challenge and most essential skill to making the PSoC useful.

Oscillator

There is an internal 24 MHz oscillator that has a ±2.5% accuracy. Since it is an RC oscillator circuit, it is subject to drift due to voltage variations and temperature. Keeping your voltage as close to 5 volts (or 3.3 volts) as possible has the largest influence on your oscillator accuracy. The second largest factor is keeping the temperature within a moderate range. This 24 MHz is divided several times within the chip, allowing you to run the CPU at various different clock rates. Choosing the highest clock speed isn't always the right answer, since a lower clock speed translates into power savings and a wider voltage operating range. Please note that the 25xxx/26xxx parts had a Flash fetch issue. Certain instructions at certain memory boundaries had the potential of being lost in the fetch from Flash. Cypress added a protection in the PSoC Designer to warn you of this situation. They also built a tool into the linker that will allow PSoC Designer to shift the Flash contents of your program around so that these instructions will not fall on those Flash boundaries and be a potential problem. These options can be selected by choosing **Settings...** from the **Project** menu and then selecting the **Linker** tab (See Figure 2-1). If your project doesn't need to run at 24 MHz, then you won't have to worry about the 24 MHz alignment errors. If you do need to run at 24 MHz, then the **Enable silicon errata warnings** checkbox will display any instance of this error potential within your code. If you select the **Enable 24 MHz alignment shift** checkbox then PSoC Designer will attempt to shift the Flash contents to prevent this situation. If your project doesn't have enough free room available, then you will need to reorganize your code until PSoC Designer is able to determine an alignment that will work. The other option is to optimize the code that is being used to free up some extra space to allow the shift to happen. Some optimization ideas are discussed in Chapter 10 – *Project Pruning*.

The PSoC can use two I/O pins (P1[0] and P1[1]) to drive an external crystal. Note that the crystal must be referenced to Vcc and not to GND. I was once told that this was the result of an oversight more than it was an intended design characteristic, since the other micros that I've designed with all reference their crystal circuitry from GND. This idiosyncrasy introduces a little difficulty in hardware design since Vcc is at the opposite end of the chip whereas GND is right next to the crystal pins. Good design dictates that you should keep the path between Vcc and the crystal as short as possible. In-circuit programming also uses the crystal I/O pins. If you are going to use this capability, make sure to keep the programming header right next to the crystal pins. This is imperative for a noise resilient system and can help you keep the capacitance on those pins under control. Unexpected capacitance on your crystal pins will adversely affect the accuracy of your crystal circuit.

The PSoC uses 24 MHz internal main oscillator as a clocking source for the digital and switched cap analog blocks around the chip. If you are running an external crystal, you are able to enable the system's phase locked loop to gain the added accuracy for this clocking signal. For flexibility, there are two or three taps (depending on which PSoC family you are using) that divide down the 24 MHz into slower frequencies if needed. These slower frequencies may be selected as the clocking source in lieu of the 24 MHz if desired. These dividers are designated 24V1, 24V2 on the 25xxx/26xxx parts and as VC1, VC2, and VC3 on the newer 22xxx, 24xxx, and 27xxx parts. The newer parts also specify a clock source called *SysClk*. It replaces the 24 MHz source of the older parts. SysClk is selectable either as 24 MHz or the input frequency of P1[4]. If P1[4] is selected as the source for SysClk, then VC1 as described below will use the frequency on that pin rather than the 24 MHz.

24V1 (or VC1) is stored in the upper nibble of the OSC_CR1 register. Selecting 24V1 as a clock selects a frequency equal to 24 MHz divided by the value of 24V1 plus one. Since you can store 0–15 in the upper nibble of the OSC_CR1 register, this corresponds to a divisor of 1–16. 24V1, therefore, is able to provide a frequency down to 1.5 MHz.

24V2 (or VC2) operates identically to 24V1 with the exception that its source frequency is the output of 24V1 instead of 24 MHz. Its value is stored in the lower nibble of the OSC_CR1. It has the same limitations as 24V1 with a value from 0–15 that corresponds to a divisor from 1–16.

VC3 is an addition to the newer families of the PSoC. It gives a third tap to use to clock the digital and analog blocks in your project. VC3 has a divider range from 1–256. Its corresponding value of 0–255 is stored in the OSC_CR3 register. VC3 has four different sources to pull its frequency from. It can use the SysClk, VC1, VC2, or SysClk*2 as its frequency.

RAM Organization

The CPU has five core registers that are not accessible in the regular RAM address space. These registers are the stack pointer, program counter, accumulator register, index register and flags register.

The 16-bit program counter allows for 64K of addressing. This is adequate for the first couple of families of PSoC parts that only have up to 16K of Flash. Parts will be in production by the publication time of this book that will allow up to 64K.

The accumulator register (A) is a general-purpose register that is designed to be at the heart of mathematical operations, port manipulation, and for special move instructions. It's important to know that you must use the accumulator to perform certain tasks. For example, let's say that you wanted to add two registers together. I will call the registers, value1 and value2. After I have added the two registers together, then I want the sum to be stored in a register called result. Here's how I do it.

```
mov    a,[value1]   ;copy value1 to the accumulator
add    a,[value2]   ;add value2 to the accumulator
mov    [value3],a   ;store the sum in value3
```

This is a valid set of instructions and should compile okay. There are a couple of different ways to accomplish this goal, but I won't expound on that here. This method may tend to be a little frustrating for those who are used to a more complex processor that will allow all mathematical operations on all register locations. If the Cypress had a more complex instruction set, you might have been able to save one instruction by coding the following instructions. One instruction may not seem like much, but for this simple task it's a 30% savings and those types of savings can add up quickly.

```
mov    [value3],[value1] ;copy value1 to value3
add    [value3],[value2] ;value3 = sum of value1 + value2
```

The first instruction in the list is completely valid. Fortunately, Cypress made an allotment to copy from register to register directly without going through the accumulator. This was a great move, as it would be much more difficult to complete a complex project if all moves had to go through the accumulator. However, the second instruction is not valid and will cause an error. You are not able to add a register to another register value without using the accumulator.

The M8C core does provide some useful methods of performing mathematical and move operations that go above and beyond what I have seen in some other accumulator-based micros. For example, I could rewrite the addition method in the following manner:

```
mov     [value3],[value1]       ;copy value1 to value3
mov     a,[value2]              ;copy value2 to the accumulator
add     [value3],a              ;value3 = sum of value1 + value2
```

This is another perfectly valid method for adding the two numbers. Even though I am still at 3 bytes for this method, which is the same as the method used above, it shows that I'm able to have the register be the final location for the mathematical operation, which can save some heartache in more complex operations.

Certain addressing methods also require the use of the accumulator and or index register. Examples include usage of the swap instruction, or moving nonimmediate values into the control registers.

The index register (X) allows for indexed addressing. It also serves as a general-purpose register that allows more operations than a general register location. It function is similar to the accumulator as it's only accessible to instructions that address that register and is not available in the regular RAM address space. The index register also allows you to perform indirect functions within the RAM address space. In the indirect addressing method, the index register is used to point into the RAM address space. Consider the following code:

```
mov     X,3             ;X = 3
mov     [X+0],25        ;RAM location 3 = 25
mov     [X+4],35        ;RAM location 7 = 35
```

The same basic concept of indirect addressing is used in other move instructions and mathematical operations.

More concepts of these instructions and RAM usage are discussed in Chapter 8 – *PSoC Memory Management*.

The stack pointer is another CPU register only accessible by two instructions:

```
mov   X,SP     ;copy the stack pointer into the Index Register
swap  A,SP     ;swap the stack pointer with the Accumulator
```

The boot.asm file will set the stack pointer to the top of the memory allocated by your project. From there, the stack pointer will grow towards higher addresses every time a value is pushed into the stack. Care must be taken so that the stack has room to grow. Refer to Chapter 8 – *PSoC Memory Management*.

The flags register contains the selection bit for choosing between bank0 and bank1 of the control registers. It also contains a bit that indicates if the processor is executing supervisory code or user code. The more typical contents of the flags register are the carry and zero bits that indicate the results of previous logical and mathematical operations. The global interrupt enable bit is also stored in the flags register.

The flags register is only available to operations used to manipulate or test various bits of the register. These instructions are TST, AND, OR, XOR. These instructions are all that you need to test the current state, clear, set, or toggle bits within the flag register. The flags register is affected by all logical and mathematical instructions, but is unaffected by many other instructions such as CALL, RET, and JMP. The flags register is saved and restored automatically in an interrupt.

The first PSoC micros incorporate 256 bytes of data RAM available for user use and for stack operation. There are some newer parts coming into production that will have more RAM than this, but the parts were not available during the writing of this book. It is my understanding that Cypress will be implementing a paging procedure to address the additional banks of RAM. For now, however, I use the current allocation of 256 bytes of RAM. For these current parts, all RAM is addressable using 8 bits, therefore there isn't any need for paging instructions. This RAM is used for stack space as well as filling the RAM requirements for any modules you choose to use. Instructions indicate usage of the general-purpose RAM by enclosing operands in square brackets. If no brackets are present and the operand isn't a CPU register, then the compiler will assume the operand is an immediate value and will try to process it as such. Consider the following:

```
mov    A,4        ;Accumulator = 4
mov    A,[4]      ;Accumulator = value of register at address 4
```

The first instruction will move a 4 into the accumulator. The second instruction will move whatever value is in the RAM register at address 4 into the accumulator. Remember that this RAM is unaffected by the state of the bank selection bit of the flags register.

Control and configuration registers are kept in two banks of 256 registers each. The user will differentiate the register banks from data RAM with the use of square brackets prefixed with reg (for example, reg[REGISTERNAME]). Bit 0 of the flags register determines which bank of configuration registers you are addressing. Following is an example:

```
mov    A,4        ;Accumulator = 4
mov    A,[4]      ;Accumulator = value of register at address 4
mov    A,reg[4]   ;Accumulator = value of configuration register 4
```

The last instruction is an example of retrieving the value from a configuration register. If I were using a 27xxx part, this address would correspond to the PRT1DR register if I'm in bank0 or the PRT1DM0 register if I'm in bank1. (Refer to the data sheet under Register Reference.)

Chapter 8 – *PSoC Memory Management* delves more deeply into RAM methods and organization.

Supervisory ROM

The supervisory ROM is a section of Flash that is outside of the normal address space. It contains special code to boot the processor, retrieve special calibration settings such as the oscillator trim value, and also contains code to manage Flash functions. Cypress has recommended that you use the sections of code provided by them to access the supervisory ROM. There are a few hoops to jump through to execute from this area of memory and it's not for the faint of heart. The EEPROM module is an example of a module that utilizes the supervisory ROM.

Interrupt Controller

The data sheet for the 27xxx part provides a very eloquent description of the interrupt controller. "The interrupt controller provides a mechanism for a hardware resource in PSoC mixed-signal array devices to change program execution to a new address without regard to the current task being performed by the code being executed." The only attention to the current task is that the processor finishes the current instruction when the interrupt comes pending. At that time, the program counter and the flags register are saved on the stack and execution begins at the appropriate interrupt vector. At the vector location, there will be a long jump (LJMP) instruction that passes control to another section of code in your project, known as the interrupt service routine. After executing the code of the interrupt service routine, the return from interrupt (RETI) instruction will tell the processor to pop the flags register and program counter from the stack and return execution to the location in code that was interrupted when the interrupt occurred.

The vector location depends on the source of the interrupt. A listing of the vector locations and their corresponding interrupt sources is found in the data sheets under interrupt controller. You may notice that the interrupt vector table has changed somewhat from the 25xxx/26xxx parts to the newer parts. The newer parts have two additional interrupt sources, VC3 and I²C. Cypress, however, has also rearranged the other interrupt sources. This may be of particular concern if you are migrating from older parts to newer parts. The interrupt vector table of the newer parts also seems to require more space than just an extra two vector locations. The boot.asm file notes that Cypress is planning ahead for future parts. By publishing time there will be another Cypress family that includes more configurable blocks and therefore more interrupt sources.

Kudos to the writer of the data sheet for the 27xxx parts as it contains, in the interrupt controller section, the most informative and clear description of interrupt processing that I've seen in a microcontroller data sheet.

It is possible to nest interrupts within the M8C core. Once an interrupt is processed, the flags register is pushed on the stack. The global interrupt enable bit is cleared after that. This prevents any additional interrupts from occurring during the processing of the interrupt. Once the interrupt service routine has finished executing and a RETI instruction is performed, the flags register is retrieved off of the stack. This re-enables the global interrupt enable bit. If you re-enable the global interrupt enable bit within the interrupt, the other interrupts are now able to occur *before* the first interrupt has finished processing its interrupt service routine. This can be very useful for critical tasks that must be serviced as quickly as possible and can't simply wait until all other interrupts are processed. You must add the necessary code for context switching. In other words, it's your responsibility to save enough information in the interrupts to maintain operation integrity if both interrupts use identical resources.

General-Purpose I/O

The general-purpose I/O has a slightly different structure compared with other micros that I have used. It not only gives you the opportunity to have a strong drive or a high impedance state, but it allows you have a resistive state in both the pull-up and pull-down configurations. Cypress has also chosen to have both the output control and the input reads of a port be handled with the same data register (PRTxDR). This architecture has some advantages, but also requires you to be careful in particular situations, or it could be the source of some hard to find bugs. First let's examine the diagram for the output and input drives given in the PSoC data sheet for the 26xxx series. The diagrams are more descriptive of the circuitry around the port pin. After looking at this structure, we'll look closer at what's different in the 27xxx series.

Output State

There are three components to determine the digital output state of the PSoC. The first two are listed as the drive mode bits and are designated DM1 and DM0. They are defined in the DM1 and DM0 registers for each particular port. These bits determine which of the logic AND gates A, B, C, and D are turned on.

The third component is the state of the data register, assuming for the moment that the global select register has disabled the global bus from this pin. The data register

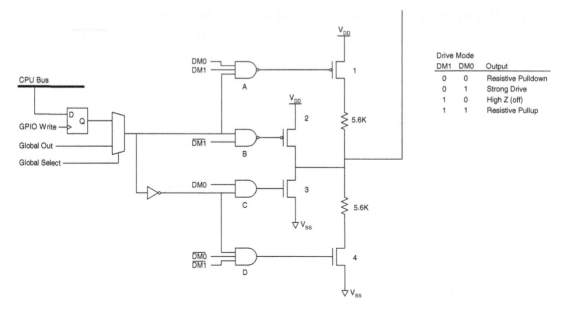

Figure 2-1: Output Port Architecture

will determine the state of the signal that is coming from the left of the figure and also feeds into the logic gates A, B, C, and D.

First we'll examine the strong drive and high impedance states, as they operate more similarly to other micros that I've dealt with. If you choose the **High-Z** (off state) in PSoC Designer, then you are setting DM1 to a 1 and DM0 to a 0. This disables all four AND gates A, B, C, and D. Therefore, regardless of the state of the data register bit, the logic switches 1, 2, 3, and 4 will always be off. The state of the bonded I/O pin will not see any current path to or from the output section of the I/O pin. This gives you your high impedance state. However, if you choose the **Strong** drive option from PSoC Designer, then you will be setting DM1 to a 0 and DM0 to a 1. This turns on AND gates B and C and turns off AND gates A and D. The state of the data register bit determines whether B or C is turned on giving you a low impedance connection to Vdd or Vss. Now you have your strong drive signal to drive components outside of the PSoC.

The pull-up and pull-down states of the PSoC are where I've seen most people get stuck, as I've had to dive in to find out why I myself was having problems. It's easy to see the confusion. PSoC Designer allows you to choose the state of the drive mode pin as a drop-down combo box from within PSoC Designer. It's easy to think, "Hey, I need a pull-up on this pin, it will save me a resistor to add to my design," and you pick **Pull-Up** from the drop-down box and cheerfully generate the application and compile the code. Yet as you power up your project, nothing seems to be working right. It's as if the pull-up resistor isn't there. What has really happened is that the data register hasn't been asserted correctly or it has been overwritten. Let's look now at the resistive states of the I/O pins.

The default power-up state for the drive mode registers for the 26xxx part is 0x00. That means that the chip defaults to the resistive pull-down state. A close look at the logic of the system will show that having a 0 for both DM1 and DM0 will turn on gates B or D depending on the state of the data register. If the data register state is low, then the D gate will turn on. This enables the logic switch thereby connecting the output pin to a 5.6K pull-down resistor. However, if the data register state is high, then the D gate will turn off and the B gate will turn on. This turns on logic switch 2 which is a strong (low impedance) connection to Vdd. Therefore, if you want to have a resistive pull-down state, then you need to make sure that you have a 0 value in the associated bit of the data register. If you have a 1 in the associated bit of the data register, then you will have a strong drive to Vdd.

Now you can start to see the problem that was encountered in the pull-up situation described previously. If you select the **Pull-up** option from PSoC Designer, then you are enabling gates A and C. This will give you the resistive pull up or it will give you the strong drive to ground depending on the state of the data register bit. The data register defaults to a value of 0x00 so unless you have changed this state of the desired bit to a 1, you will not have the resistive pull up you were looking for. You will have a strong drive to 0 and your project will appear to be broken. I would not be surprised if the newer version of PSoC Designer will include code in boot.asm to assert the data register value to default you to a pull-up state if that is what you have selected from the Device Editor of PSoC Designer, but version 4.1 does not seem to do it. Even if/when they do, you will need to make sure that this bit stays high in order to maintain the pull-up state.

Input State

Understanding the input section of the general-purpose I/O of the PSoC family is important to prevent those difficult to find bugs that come and go. Examine the diagram of the input section of the general-purpose I/O of the 26xxx family in Figure 2-2.

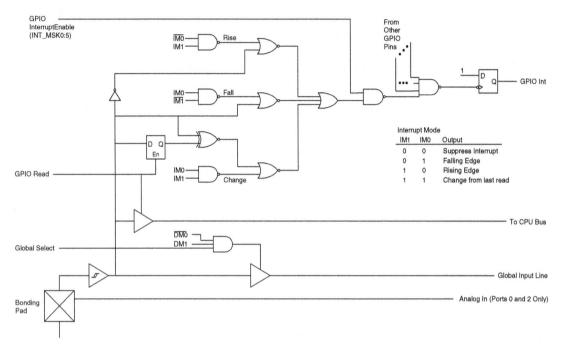

Figure 2-2: Input Port Architecture

First and foremost to notice about the input section is that the digital read of an I/O pin is directly from the bonding pad. This is referred to in the diagram as the line *To CPU Bus*. Since this pin comes directly from the bonding pad, it reflects the actual state of the pin regardless of the signal that you are outputting to the pin. Since the output stage controls whether or not the pull-up or pull-down state is enabled, you may not be able to completely disable the output circuitry when reading a pin.

This is useful since you are able to see if the output state is being overloaded by the outside world. An example of this would be using the pull-up resistive state to assert the state of a normally open momentary switch. When the switch is closed, it shorts the bottom pull-up resistor to Vss. Since the input circuitry reads the state of a pin and not the high output signal needed to assert the pull-up resistive state, you are able to read the state of the button. In a more atypical operation, you would be able to read if the output circuitry isn't enough to drive a high current load because you would be able to read the state of the bonding pad.

This presents a trap for instructions such as AND, OR, and XOR. These instructions require a read of the port before writing out to the port. Since the read of the PORT isn't necessarily what is written out to the port, you can end up with a port configured incorrectly. Let's look at a hypothetical situation. Let's say that PORT 1 bit 0 has an LED connected to it in series with a resistor. Bits 1, 2, and 3 are connected to momentary switches that short their respective pins to Vss when activated. Bits 4–7 aren't connected on this project. I've illustrated this with a simple schematic, Figure 2-3.

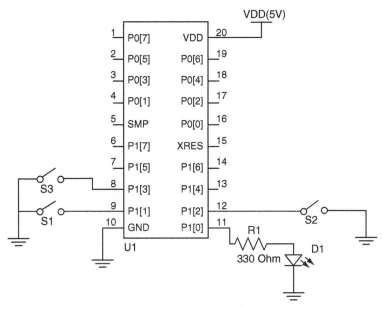

Figure 2-3: Schematic

In my project, the switches are used as a user interface that initiates certain tasks. My routine might look something like the following:

```
check_tasks:
    tst    REG[PRT1DR],0x02   ;Check to see if first button is pressed
    jz     task1              ;Jump to task 1 if pressed
    tst    REG[PRT1DR],0x04   ;Check to see if second button is pressed
    jz     task2              ;Jump to task 2 if pressed
    tst    REG[PRT1DR],0x08   ;Check to see if third button is pressed
    jz     task3              ;Jump to task 3 if pressed
notask:                       ;Exit if no button is pressed
    ret
task1:
    ;Perform task1 here
    ret
task2:
    ;Perform task2 here
    ret
task3:
    :Perform task3 here
    ret
```

This section of code doesn't pose the risk that comes from using the AND, OR, or XOR instructions since TST doesn't write back out to the register PRT1DR. However, let's look at what happens when we write a small routine to toggle the state of the LED.

```
toggle_led:
    xor    REG[PRT1DR],0x01        ;toggle the state of the LED
    ret
```

Let's say that the current state of bit 0 of port 1 before calling this routine is a low. Therefore a binary view of the states of each bit of port1 would be as follows:

Bit	7	6	5	4	3	2	1	0
State	X	X	X	X	1	1	1	0

Since bits 4–7 aren't connected to anything, we won't worry about them for the moment. The LED state is 0, but each of the buttons has a state of 1 (high). Remember that I have to write a 1 to each bit in order for the pull-up state to work.

Now when the toggle_led routine is called, then you expect to affect only the LED as shown below. I'm going to assume a value of 0 for the don't care bits.

```
        0000 1110
XOR     0000 0001
        0000 1111
```

That seems to work as the LED has been toggled appropriately. However, in reality, it will only work this way if none of the switches have been closed! Remember that the PSoC reads the actual state of the pin and not the output register value, therefore, if someone were pushing the switch that is tied to bit 1 of port 1 then my equation would end up with a different answer.

```
        0000 1100
XOR     0000 0001
        0000 1101
```

Since I read a 0 for the value of bit 1, I end up writing 0x0D back out to the port instead of 0x0F. Recalling the structure of the previous output port, this will then cause the pull up to be disabled and a strong low to be sent to the bonding pad. From that read on, I will always read a 0 at bit 1. The same type of problem can happen if you have configured an I/O pin to have a pull-down configuration.

Depending on how the bit is used in your code, this could introduce a difficult bug. This can be even harder to find if you write the proper value out to that pin at some unknown time in the future, thereby correcting the problem. Then you have a bug that comes and goes. There are a couple of different approaches to preventing this problem. Cypress uses the pull-up and pull-down configuration in some of their module development. Their solution is to use a shadow register.

The shadow register is a copy of what should be loaded out to the desired port register. In the previous example, the shadow register would be loaded with the value of 0x0E during the initialization.

```
toggle_led:
   xor  [shadow_port1],0x01   ;Toggle the LED bit in the shadow register
   mov  A,[shadow_port1]      ;Move to accumulator to move to port
   mov  reg[PRT1DR],A         ;Set new value out to port 1
   ret                        ;Return from this routine
```

The shadow port works well, but you need to be aware when it is used as you need to change the shadow register every time you change a port pin value. This became painfully apparent when I tried to use the LCD Module in PSoC Designer. They utilized a shadow register on the port to which the LCD is connected. They don't use every pin of a particular port, so they implement a shadow register so as not to change the port register value for the unused pin inadvertently. Every time I tried to change the value of that extra pin, I was very confused because that value would seemingly change back on its own shortly after. It took a little while to track down that the pin was reverting to its original value every time that I called the routine to update my LCD display. The LCD routine writes the shadow register back out to the port and thereby overrides my value at the extra port bit. In order to change the port bit effectively, I need to change both the shadow register and the port bit, so that the next time I update the LCD, that routine is maintaining the desired value at the port.

If you don't want to devote an extra byte of RAM to function as a shadow register, you can assert the necessary bits as you change the port value. If there are many different locations where you write to the port, you will probably want to make this section a subroutine or a macro. This would change your toggle_led routine as follows.

```
toggle_led:
   mov   A,reg[PRT1DR]   ;Get current value of port 1
   or    A,0x0E          ;Make sure all pull up bits are high
   xor   A,0x01          ;Toggle the LED bit
   mov   reg[PRT1DR],A   ;Set new value out to port 1
   ret                   ;Return from this routine
```

That will take care of setting the bits as needed for our particular project. If you were utilizing the unused pins in the pull down configuration, then you would need to make sure that they are cleared properly in addition to making sure that bits 1–3 are set properly. This would add one more line.

```
toggle_led:
    mov    A,reg[PRT1DR]        ;Get current value of port 1
    or     A,0x0E               ;Make sure all pull up bits are high
    and    A,0x0F               ;Make sure all pull down bits are low
    xor    A,0x01               ;Toggle the LED bit
    mov    reg[PRT1DR],A        ;Set new value out to port 1
    ret                         ;Return from this routine
```

That's not too painful and you save a byte of RAM for other uses. Remember that you need to look at this concern every time you implement a pull down or pull up configuration on your port. You will also need to be aware if the module you have chosen utilizes a shadow register for the port.

The next thing to notice about the input structure of the PSoC I/O pin is the global input line. It is controlled by a logic gate that is switched by the global select. This is needed so that there isn't any contention on the global bus as the same bit of every port will have a connection to this same line. However, they also show that DM0 and DM1 play a role in the state of that logic gate. They intend for the port pin to be in a High-Z configuration. When you choose global in as the drive mode for a pin in Device Editor, then PSoC Designer will compile the boot process so that it will turn on this gate before going to the main.asm file. However, if you should try to change these registers manually so as to implement a pull-up or pull-down configuration, then you will disable this gate. Therefore, all pins that are going to work as a global signal will need to implement external pull-up or pull-down resistors. Note that you can still read the logical state of a pin via the data register for that pin even though it is configured as a global input.

The final note that I want to point out about the input structure of the I/O pin of the PSoC is the interrupt section. You will notice from Figure 2-2 that the interrupt section utilizes interrupt mode bits IM0 and IM1 to determine whether a particular pin is disabled, or set up to trigger on a rising edge, falling edge, or either edge. These interrupt mode signals are unique for every bit on every port. However, notice that the general-purpose input/output (GPIO) interrupt enable bit is the same for every GPIO pin, and the interrupt signals from all other GPIO pins join in at the last stage before tripping the interrupt. There is only one interrupt vector for all pins that are configured as a standard CPU IO pin function. It is left to the user to determine which GPIO tripped the interrupt when it occurs. This can be extremely difficult for a fast changing signal on the pin triggering the interrupt. It could lead to confusion on how to handle the interrupt.

You should also note that since the last stage of the interrupt logic that combines the other GPIO interrupt logic in with this one is simply an OR gate, you will not be able to distinguish when two GPIO interrupts occur at the same time. It would appear the same as one GPIO interrupt.

The newer PSoC parts such as the 27xxx series implement another drive mode bit (DM2) to control additional circuitry in the drive logic at the output stage of the GPIO pins. This additional bit implements the following modes:

1. Open Drain, Drives High
2. Slow Strong Drive
3. High Impedance Analog
4. Open Drain Drives Low

The exact details of how this circuitry is implemented aren't included in the data sheet. You can probably picture how it would be accomplished. The open drain methods are a great addition to help with such situations as an I^2C buses. The slow strong drive adds additional slew to the output signal at the bonding pad. This has helped me in projects where I was getting a little too much noise emission from my circuit in an audio application. Setting the slow strong drive option on the ports reduced the audible noise in my circuit dramatically. The last option, high impedance analog, shuts off the digital reading of the input pin. I believe that it might also power down some sections of that circuitry as I'm told that choosing this option is a bit of a power savings. Note, however, that if you choose this option, you will not be able to read the digital input of the pin correctly.

Analog I/O

The analog capabilities of the Cypress part really do set it apart from most microcontrollers. I've had a lot of fun working with the different possibilities that the analog I/O and associated blocks offer. The PSoC allows connections into the analog section of the micro via the continuous analog blocks. They are the top row of analog blocks as viewed in Device Editor. The newer families also allow a direct connection into certain switched cap blocks.

The analog inputs and outputs are associated with port 0. You should reserve pins from port 0 in for your analog functions. The input connections to the continuous analog blocks are chosen using analog multiplexers as shown in Figure 2-4.

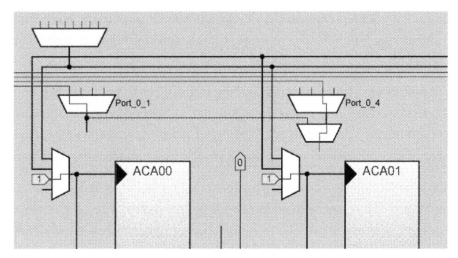

Figure 2-4: Input Muxes

The multiplexer above block ACA00 is referred to as AnalogColumn_InputMUX_0. The multiplexer above block ACA01 is referred to as AnalogColumn_InputMUX_1 and so on. The newer 27xxx and other parts have the same structure for input multiplexers. You can consider the connection to the following listed ACAxx blocks to be the same for the ACBxx blocks of the newer parts.

As you use your mouse to select each multiplexer, you will notice that AnalogColumn_InputMUX_0 only allows four possible connections. These are Port_0_1, Port_0_3, Port_0_5, and Port_0_7 or the odd bits of port 0. AnalogColumn_InputMUX_1 allows connections to the even bits of port 0. There is also a multiplexer under AnalogColumn_InputMUX_1 called AnalogColumn_InputSelect_1. This multiplexer allows an additional option to select the source signal for ACA01 to be able to come from either AnalogColumn_InputMUX_0 or AnalogColumn_InputMUX_1. ACA00 will not be able to connect to even bits from port 0 as there isn't a way to make that connection. Likewise odd bits from port 0 are not able to connect to the ACA03 block.

Once the analog signal is routed to a continuous analog block, it can be routed to other analog blocks within the PSoC so it can be routed back out of the chip via the analog output buffers. There is one output buffer per column of analog. The buffers are only able to connect to one pin each. These pins are set and cannot be changed, so they should be reserved early in the design after block placement has been finished. The analog buffer output pins are, from leftmost column to right, P0[5], P0[3], P0[4], P0[2].

The output buffers are also continuous and are not subject to switched cap limitations, unless, of course, you are outputting the signal of the switched cap block through the analog output buffers. They can only output what is fed in, so you will see the switching characteristics of the switched cap analog blocks, such as frequency limitations.

The typical current flow associated with the analog input of the 25xxx/26xxx parts is 3 nA (nanoamps). That has been adequate for my projects not to drag down an input signal. However, the newer parts boast a noticeable improvement of only drawing 1 nA. The maximum rating is also much lower on the newer parts.

Usage of the older parts is also subject to a rail-to-rail limitation that should be carefully researched. It is around 0.5 volts on both the bottom and top rails. I would recommend that you use the newer 'enhanced analog' parts on all new designs if possible. Refer to Chapter 4 – *Limitations of the PSoC* for more information on some of the analog limitations of the older parts. These limitations apply to both the analog inputs and analog outputs of the PSoC. The data sheet also mentions that the same analog limitations apply to the operational amplifier at the heart of the switched cap blocks. However, I haven't noticed the same difficulty with reading voltages close to the rail with the analog-to-digital converters on those blocks, so the structure of the block may hide it better. That data sheet of the SAR6, for example, doesn't mention the limitation and says that its input range is rail-to-rail. I don't understand why this is the case, but the SAR6 seems to work fine for me.

Since the older parts don't allow for a rail-to-rail signal to be processed through the continuous analog blocks to the switched cap blocks, you are able to implement the test mux that ties the input voltage directly to the output voltage to bypass this limitation. This also means that you aren't able to amplify or attenuate the signal as you have defeated the op-amp in the process. There is some amplification that you can employ through use of the switched cap blocks to help out your analog-to-digital conversions if needed. Another method is to use an analog-to-digital converter that has more resolution.

You should notice that the output buffers have an output impedance of around 1 Ohm listed in the electrical specifications. This, coupled with the ability to use around a milliamp or so of current on the older parts and a couple of milliamps of current on the newer parts, dictate that the output buffers are intended to provide a signal and not drive high-power circuits directly.

Digital and Analog Programmable Blocks

The digital and analog blocks are modules of analog and digital hardware that can be configured for different types of operation via a set of registers specific to each block. The different blocks are separated into groups that have identical functions and capabilities. These groups are as follows for the 25xxx/26xxx parts: Digital Basic Type A, Digital Communication Type A, Continuous Analog, Switched Capacitor Analog Type A, and Switched Capacitor Analog Type B. The newer parts have blocks that are improved and therefore have been given different designations. A comparison is given in Table 2-1. The various blocks are listed in order from left to right, top to bottom as they appear in the Device Editor, Interconnect View, of PSoC Designer.

Table 2-1

25xxx/26xxx	22xxx/24xxx/27xxx/29xxx
DBA00	DBB00
DBA01	DBB01
DBA02	DCB02
DBA03	DCB03
DCA04	DBB10
DCA05	DBB11
DCA06	DCB12
DCA07	DCB13
ACA00	ACB00
ACA01	ACB01
ACA02	ACB02

Table 2-1 (continued)

25xxx/26xxx	22xxx/24xxx/27xxx/29xxx
ACA03	ACB03
ASA10	ASC10
ASB11	ASD11
ASA12	ASC12
ASB13	ASD13
ASB20	ASD20
ASA21	ASC21
ASB22	ASD22
ASA23	ASC23

Those blocks that begin with 'DB' are basic digital blocks. There are a few differences in how the block can interconnect on the newer parts, so the newer parts list these as 'DBB' rather than 'DBA'. The digital blocks that begin with 'DC' are digital communication blocks. These have some additional functions that aren't found in the basic digital blocks. There still are some differences in interconnects, so the newer parts specify their digital communication blocks as 'DCB' where the older parts are 'DCA'.

You should also note the difference in order for the placement of the basic digital blocks and the digital communication blocks. This order of placement is significant on how the clocking signals for the digital communication blocks can be sourced. Also the additional interconnection opportunities for the newer parts allow you to connect inputs and outputs of the digital communication blocks to every pin on the chip. This was not possible with the older PSoC devices. Chapter 4 – *Limitations of the PSoC* has more details of these differences.

The blocks that begin with 'AC' are continuous analog blocks. The added features of the newer parts prompted the new designation of 'ACB' rather than the 'ACA' of the older parts. Likewise the differences of the switched cap blocks requires a new set of designations 'ASC' and 'ASD' where before the switched cap blocks were denoted 'ASA' and 'ASB'. The blocks of the newer parts are very similar in their functions, but the newer blocks have some added enhancements that make them more powerful.

Basic Digital Blocks

The basic digital blocks are groups of logic that are able to perform a variety of functions. They can be operated independently or in concert to form timers, counters, PWM generators, random number generators, or cyclic redundancy check calculators. The function and associated configurations are controlled via a set of registers that are specific to each individual digital block.

It's important to note that the bits of the configuration registers are interlinked. A particular bit may have a different effect on the function for different values of the configuration registers. Care is taken in the configuration files generated by PSoC Designer to load all configuration registers in an order that avoids unexpected glitches in operation. If your project requires changing multiple registers, make sure that the order of changing the registers is considered carefully.

The various input and output signals required for the digital blocks can be derived from input and output signals of other blocks or input/output pins of the chip. The blocks are also able to connect to internal clocking signals within the chip. This allows the large number of possible configurations that Cypress touts.

Digital Communication Type A Blocks

These digital blocks are similar to the basic digital blocks and are able to perform all the functions of the basic digital blocks with the addition of being able to operate as an SPI master or slave, or as a UART transmit or receive. These blocks are the bottom row of digital blocks on the 25xxx/26xxx parts or are the rightmost two blocks on both rows of digital blocks on newer parts such as the 24xxx/27xxx/29xxx parts.

When operating as a UART transmit or receive block, it is necessary to provide a clocking signal to the block in order to generate the proper baud rate for the UART. One single block cannot serve both functions of clock generation and a UART transmitter or receiver. This clock signal can come from any of the typical clocking signals available to that digital block such as the global clocking signals, a global bus line, or a signal output from another block.

If you are going to operate as an SPI master, older versions would require you to use a general-purpose I/O pin and control your own slave select signal. The newer PSoC modules have included that function within the SPI module, but if you want to deviate from their operation of the slave select signal for any reason, then you may want to continue to use your own general-purpose I/O pin.

Continuous Analog Blocks

The continuous analog blocks are the top row of analog blocks and can be configured to perform gain functions on analog signals. They can amplify, attenuate and invert input signals that are within their range of operation. You can connect two of them together to get a differential amplifier, or they can be set up as a comparator with a programmable threshold level. As noted in the table above the older PSoC parts have designated their continuous analog blocks as type A and the newer PSoC parts have designated their continuous analog blocks as type B. I have included graphical descriptions of these analog blocks below.

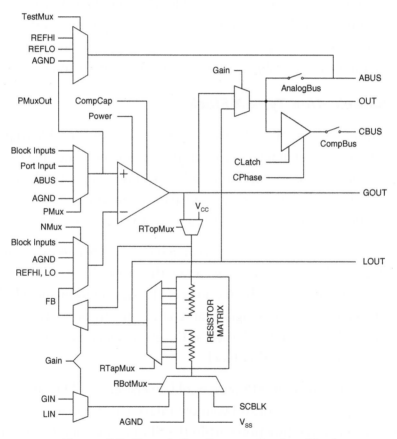

Figure 2-5: Type A Continuous Analog Block

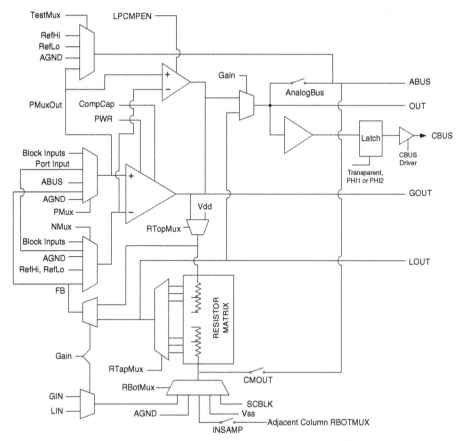

Figure 2-6: Type B Continuous Analog Block

The continuous analog blocks are able to connect to the analog input pins of PORT 0. They are also able to connect to adjacent continuous analog blocks allowing you to daisy chain gain stages together. Outputs of the continuous analog blocks can be sent to switched cap blocks for further processing of the signal or sent on the analog output bus to the analog output buffers to external output pins of the PSoC.

The comparator output signal can be used as a source for some digital functions in the digital blocks. The comparator output signal can also serve as an interrupt source.

Improvements have been made in the operation of the continuous time block on the newer parts. These improvements include rail-to-rail capability and improved accuracy on the actual and calculated gain. Chapter 5 – *Improvements of the PSoC* will explain some of these improvements in greater detail.

Switched Cap Blocks

Switched cap blocks start out with the same op-amp that is used in the continuous analog blocks. This op-amp then has various capacitors linked with switching signals placed in feedback, input, and output positions so as to serve various functions such as integrators, summers and filters. These capacitors are linked to timed switching signals. Together, they have an ability to function the same as a resistor would in the circuit. The value of the capacitance along with the speed of the switching allows the user to vary the correlating resistance value. This is the heart of the switched capacitor technology. Since switching is involved, the output signal that you would expect isn't continuously present and must be sampled at specific times in order to see the correct signal. The construction of the PSoC switched capacitor blocks allows you to process these signals easily in filters and put them out to the analog output buffers or use the blocks as analog-to-digital converters without being a switched capacitor expert. For those who are really adventurous, Cypress has created the generic switched cap block. This block gives you a visual configuration of all the settings of a switched capacitor block to allow you to create your own module easily.

There are two types of switched capacitor blocks on any particular PSoC part. On the older parts 25xxx/26xxx, we noted that these blocks have an 'A' or a 'B' designation. On the newer parts such as the 27xxx, these blocks have a 'C' or 'D' designation. The new references are due to some improvements in the switched cap blocks and are used to differentiate these blocks from the older technology.

I've included the graphical representations of these blocks.

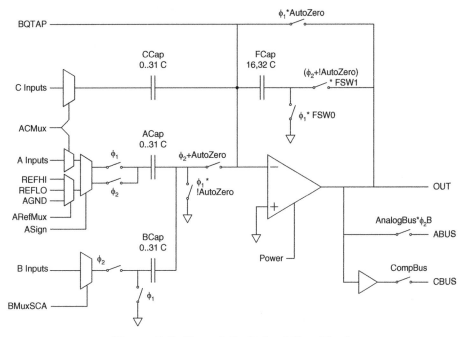

Figure 2-7: Type A Switched Cap Block

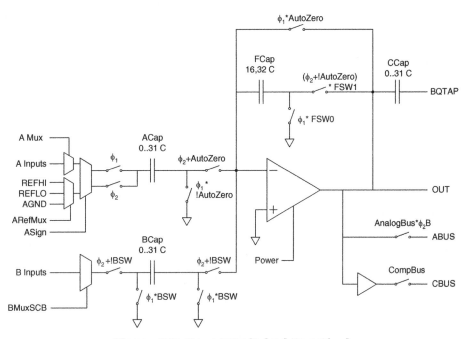

Figure 2-8: Type B Switched Cap Block

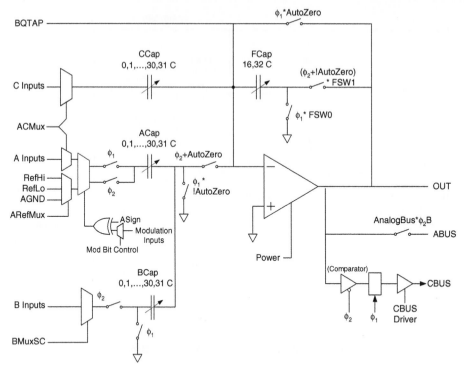

Figure 2-9: Type C Switched Cap Block

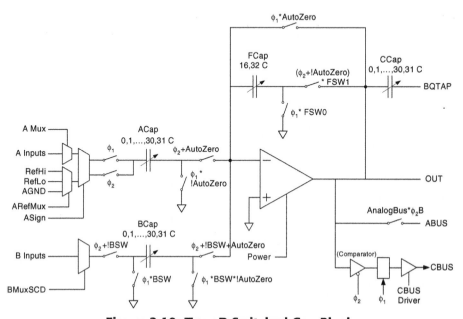

Figure 2-10: Type D Switched Cap Block

The graphical descriptions for the older switched cap blocks and the newer switched cap blocks are almost identical. However, upon closer examination, you can see that there are some differences in how the ASign input is shown and the how the comparator bus in shown. I'm not aware of any actual difference in the operation of these two parts of the switched cap blocks.

There are some connection differences that are not shown in these diagrams. The type C blocks do offer more connection possibilities that aren't found in the type A blocks. Although I considered putting diagrams in to show these connection possibilities, the diagrams quickly became too complex to be useful. Rather, I recommend that you examine the control 1 register for the type A blocks and the type C blocks from the appropriate data sheets. This will allow you to see the added connections of the type C switched cap blocks. One of the notable differences allows the type C blocks to input a port I/O pin signal directly into the switched cap block. These signals must be on P2[1] or P2[2], and they can only be directly into certain blocks, but the possibility does exist. The added configurations allow for more flexibility in filter design and other switched cap functions. There are no connection differences between the type B and type D blocks.

Switched cap block inputs can come from continuous blocks of the same column if they are in the top row. If they are in the bottom row, then they are not able to connect to the continuous time block directly, but they can connect to the analog bus signal. This signal can be connected to the output of the continuous time block and thereby you can pipe the signal from the continuous time block to a block on the bottom row directly.

Switched cap blocks can output their signal to the analog output buffers, which allow you to output a filtered signal or use a switched cap block as a digital-to-analog converter (DAC). You can also send your signal to other blocks for further processing. Using this type of system, I can bring a analog signal into the PSoC, amplify that signal, direct the amplified signal through a bandpass filter, run the output of the filter through a analog-to-digital converter, read the value of the analog-to-digital filter to perform some digital signal processing and then output the result of the digital signal processing to a DAC signal to be output on an I/O pin to the system external to the PSoC.

The PSoC switched cap blocks are very powerful. I don't even pretend to understand all that they can do, but PSoC has made their control easy enough that I'm able to make them perform many functions with minimal effort.

PSoC Designer

One of the best ways to explore the structure of the PSoC micro is through the use of PSoC Designer. This section will acquaint you with the various features of PSoC Designer.

PSoC Designer is a great tool to assist you in configuring the digital and analog modules. It also provides an editor for your source files, and a compiler, linker and programmer. It aids you in configuring the I/O and has a built-in debugger.

PSoC Designer has been though a couple of iterations since I started designing with PSoC. This book will use version 4.1 and 4.2 (the latest releases at the time of publication) for its examples. Version 4 or later is necessary for the newer PSoC parts 27xxx and 28xxx. The latest PSoC Designer can be downloaded from the Cypress Semiconductor Corporation website (http://www.cypress.com, formerly http://www.cypressmicro.com).

PSoC Designer is divided into three main areas: Device Editor, Application Editor and Debugger.

Device Editor

The Device Editor is used to configure your part's digital and analog blocks as well as the global resources for the part. The Device Editor is divided into two sections: the user module selection view and the interconnect view.

User Module Selection View

The user module selection view (Figure 3-1) is used to select modules needed for your project. You will notice in Figure 3-1 that the title bar for PSoC Designer will show the name of the current project with the project's selected part in brackets. This information remains in the title bar regardless of what view is currently selected. The user module selection view shows a list of potential modules that you can add to your project along the left side. The modules have been arranged into groups of similar functions. The groups include analog-to-digital converters (ADCs), amplifiers, analog communications, counters, digital-to-analog converters (DACs), digital communications, filters, generic, miscellaneous digital, multiplexers (MUXs), pulse width modulation (PWM) blocks, random sequence generators, temperature, and timers. Selecting the gray title button for a group will expand the modules available included in that group below the button.

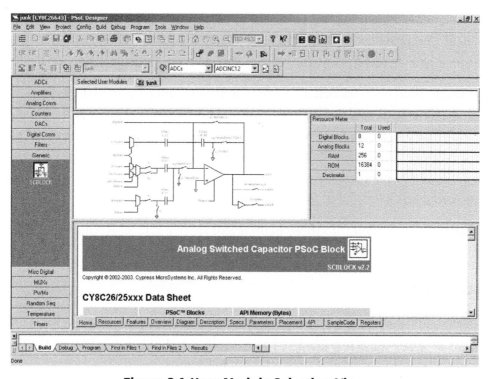

Figure 3-1 User Module Selection View

Data Sheet Section

The bottom of the user module selection view shows the data sheet for the currently selected module. Each data sheet has been divided into similar sections. Quick tabs at the bottom of the data sheet view allow you to navigate quickly to the desired section of the data sheet. Following is a quick description for each section:

Home – This is the title bar at the beginning of the data sheet. The title bar gives the name of the currently selected module. The far right of the title shows the icon associated with this module. You'll notice that modules with similar functions may have the same icon, for example, the DAC6 and DAC8 modules. Underneath the icon for the module, you will see the version of the module. This can be important for support reasons. As Cypress Semiconductor Corporation finds bugs or add improvements to user modules, they will provide updates for these modules on their web site (http://www.cypress.com). Examining the data sheet will show you which version is currently installed on your system.

Below the title bar, there is a declaration of which family of parts you are working with. It's important to note that the declaration matches the part that you have selected for your project since data sheets for the same module will have differences depending on which family of parts you are dealing with.

Resources – This tab will show what resources are required by this module. The resources needed to implement the module are divided into sections for easier reading using a table format. The PSoC blocks usage is divided into digital, analog continuous blocks, and analog switched capacitor blocks. There is also a section that shows application programming interface (API) memory usage. This section is subdivided into Flash usage and RAM usage. Please note that these numbers show what will be used in your project upon generating the application files. You can modify the files after they are generated to eliminate unnecessary code that you may want to trim out when searching to save a few bytes.

> **Note: The table doesn't give you specific requirements about block particulars, but is to be used as a general reference. For example, an 8-bit serial receiver block is shown as needing one digital block. However, in my project, this module will require one of the digital communication blocks which limits its placement into four of the available blocks.**

Features, Overview, Diagram – Gives a quick overview of your part's functions and capabilities.

Description, Specs, Timing, Parameters – These sections give a more detailed description of the construction of the module including electrical specifications and timing associated with the module. These characteristics may change from one family of parts to another. For example, the programmable amplifier gain module in the 25xxx/26xxx part family has an input range limitation that is different in the 24xxx/27xxx family.

API, SampleCode – These sections will give you a description of the subroutines created by the application code generator to allow you to start, configure, and make use of the modules that you've selected. The sample code is given in both assembly and in C for your convenience.

Registers – This section is a table format showing which registers within the PSoC micro are used to configure and control the module. These registers can be accessed in two ways when you are writing code. They can be accessed by the standard PSoC name, such as ASA01CR1, or by using the module's associated name such as DAC6_1_CR1 (the default name for a 6-bit DAC module).

Resource Meter

The right side of the user module selection view displays the resource meter. This meter is used to indicate what resources are used in your part due to configuration. The **Total** column indicates what resources are available based on the part selected for your project.

The **Used** column indicates what resources have been used by this configuration, based upon what modules you have chosen for this configuration.

The **Used** column isn't always reflective of what exactly is used in your project as the resources aren't really gone until you place the modules as described in the next section of this chapter. There are also some code compilation optimizations that can be performed that will help to save on usage of ROM. The meter is intended as a reference only.

Module Diagram

Directly above the data sheet is a diagram of the module. It is given as a quick visual reference to the operation and design of the currently selected module.

Selected User Modules and Configurations

This section shows which configuration you are currently in. Your project will create one configuration for you. The name of your project is given to this configuration and it becomes the default configuration for that project. The configuration names will show up as tab markers near the center of the screen just below the toolbar icons.

A double-click on a module from the left will add that module to the project. Alternatively, you can right-click on the module and choose **Select** to add it to your project. The resource meter will reflect the added resources that will be used when you place the module and the module will appear in the top center of the user module selection view with a default name for that particular instance.

You can rename the module with a name of your choosing. Even though you might be able to keep all the inner workings of your project straight, this is useful particularly for the next person who has to work with your project. Renaming the module is accomplished by selecting the instance that you want to rename and then right-clicking on that instance. One of the selections of the **Context** menu will be **Rename**. Select the **Rename** option and you can type in a new name for this module. The name can't have a space in it, and only allows a limited type and number of characters, very similar to a filename. If you have selected an improper name, PSoC Designer will stop you as soon as you press enter.

Additional configurations can be added by selecting **Loadable configurations** → **New** from the **Config** menu. More details on managing multiple configurations can be found in Chapter 9 – *Multiple Configurations*.

Interconnect View

The interconnect view allows you to set up global resources, place and configure modules, set up interconnections between blocks and I/O, and configure I/O pins. Refer to Figure 3-2.

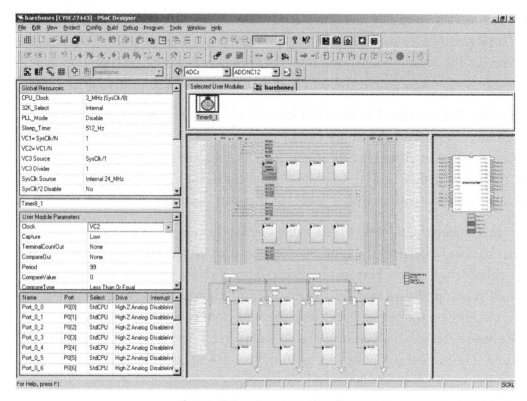

Figure 3-2 Interconnect View

The global resources are found in the top left of the interconnect view. You will notice a drop-down box next to the name of each global resource. The drop-down box will list all possible settings for that resource. By choosing a particular setting for the global resource, you are telling PSoC Designer what value to load in the associated control register during boot.asm. Following are quick descriptions of global resources. Refer to Appendix A for more information on global resources.

CPU Clock

The CPU clock sets the frequency of the instruction clock to the M8C core. This clock is based off SysClk. (Note: SysClk is the internal 24 MHz clock on 25xxx/26xxx parts.) You can choose one of several divisions of this clock if you want a slower operation frequency. Before you start thinking that faster is always better, take into account that a slower processor uses less power and is more noise immune. Make sure that you take your power supply into account, as the slower speeds allow you to operate the processor over a larger voltage range without risking bad fetches from Flash memory. The 24 MHz speed may also require some adjustments to your project if you are using 25xxx/26xxx parts.

32K_Select, PLL_Mode

The 32K_Select and PLL_Mode refer to the use of an external 32.768 kHz crystal on pins P1[0] and P1[1]. If you select **Internal** then the micro assumes no external crystal is present to use and these two I/O pins are available for normal use. If you choose **External**, then the micro tries to drive a crystal at those pins. The PLL_Mode option allows you to phase lock the internal oscillator with the external crystal. This gives you a higher accuracy on your internal 24 MHz.

Sleep Timer

The sleep timer resource allows you to adjust the frequency of the sleep timer. The sleep timer allows the processor to periodically awake from a low-power state to do a quick poll on any desired inputs or conditions to see if it's time to exit sleep or just time to go back to sleep. A slower sleep period saves power, but also means that you will have a slower response waking up from the lower power state.

VC1, VC2, VC3 (24V1 and 24V2 on Older Parts)

The VC clocks are available to use as timing sources for digital blocks, and switched cap clocking signals. The VC1 (24V1) is derived from SysClk and has the potential to divide by integer values up to 16. VC2 (24V2) is derived from VC1 and has the potential to divide by interger values up to 16. VC3 (only present on newer parts) has four sources: VC1, VC2, SysClk, or SysClk*2. VC3 has the potential to divide by 256. VC3 also can be an interrupt to the processor on terminal count.

SysClk Source, SysClk*2 Disable

These resources don't exist on the 25xxx/26xxx parts. It allows you to select a source for your SysClk. It can be the internal 24 MHz oscillator or an external source. If SysClk*2 isn't used in your project, you have the ability to disable it.

Analog Power

This allows you to turn your switched cap blocks on and off. It also allows you to set a power level for the reference voltage. The default setting for this resource is SC On/ Ref Low, meaning the switched cap blocks are on and the reference voltage is at a low setting.

Ref Mux

The ref mux can seem a little confusing. For beginning users of the PSoC I would recommend having this setting at Vcc/2 ± Vcc/2. This allows your analog-to-digital converters to work from GND to Vcc and your gain stages to work as you would probably expect. The ref mux sets the reference levels that are used in the analog blocks of the PSoC. There are also some options to use external pins to set these levels.

AGND Bypass

Enabling the AGND bypass in conjunction with placing a capacitor from Port2[4] to GND helps to reduce noise levels on the internal AGND signal of the PSoC.

Op-Amp Bias

The op-amp bias is left low for most projects. A high setting gives you a faster slew rate, but less voltage swing.

A_Buff_Power

This is a power setting for the analog output buffers. Choices are low or high. High power can mean a more stable voltage.

Switch Mode Pump

This option allows you to control whether the switch mode pump is enabled or disabled on the PSoC. The switch mode pump is used with a simple circuit to pump up a low voltage source to a voltage level where the PSoC can operate.

Trip Voltage [LVD(SMP)]

The trip voltage allows you to set two levels. LVD is the level at which a low voltage detection circuit will trip. This is similar to a brown out voltage level that you've seen on other microcontrollers. The SMP is the level at which the switch mode pump starts to transition to pump the voltage back up to a safe level.

LVD ThrottleBack

LVD ThrottleBack doesn't exist on the 25xxx/26xxx parts. It is designed to slow down processor operation at lower voltages to help with accurate operation.

Supply Voltage

The supply voltage option allows you to select between the two expected operating voltages. This option is used to determine oscillator trim values for the internal oscillator.

Watchdog Enable

The watchdog enable option is used in boot.asm generation to decide whether the watchdog timer should be enabled by default.

I have only touched on the basics of the global resources in this chapter. You will find a more detailed discussion on global resources in Appendix A.

User Module Parameters

The user module parameters appear directly under the global resources. These are the parameters that are associated with the currently selected module and are used to determine the values loaded into that block's registers at boot up. The values available for this module will appear as a drop-down box or can be adjusted with buttons to the right of the value. This gives you a visual method to see the available values for that particular parameter.

I/O Pin Assignments

Under the user module parameters, you are able to control the state loaded into the I/O pins at boot up. You are able to give a name to the port pins, used during the generation of the project's data sheet. The name is not currently used in code generation.

You are also able to select the type of I/O that the pin will be, whether it is controlled by the port's data register or if it is intended to be connected to a global signal. You are also able to set the drive mode for the I/O pin as is described earlier in this chapter.

The final setting for the I/O pin is the interrupt setting. Options are disabled, rising edge, falling edge, or a change from last read.

Selected User Modules

The user modules that have been selected for the current configuration appear at the top center of the interconnect view. If there is more than one configuration, then you will see other configurations available as tabs at the top of the selected user modules section. A right-click on the module in this section allows you to place or unplace the module, call up the data sheet, block diagram, adjust the color, or rename the module.

Blocks and Interconnects

The center of the screen holds so much information that efficient use of the pan and zoom functions are needed to see what is really going on. A click of the mouse while holding the **Ctrl** key will zoom in on the area where the mouse is located. A **Ctrl + Shift** with a mouse click will zoom back out. Holding the **Alt** key will dragging with the mouse allows you to pan around different areas of the screen.

Certain parameters of the digital and analog blocks can be set by clicking the respective graphical location of the module placed in that block. You are not able to set parameters of empty blocks in this manner.

Left clicking on a multiplexer, analog output buffer or analog clock will allow you to choose the options available for that particular resource. Improvements to PSoC Designer when they created 4.x have made the connections and options much more visible than they were in earlier versions. You should be able to see visually a connection from a digital block outputting a PWM signal go from the digital block through the lines on the screen to the icon with the name of the desired pin. This is one area where you just need to spend some time clicking around on the screen to discover what is possible.

IC Graphic

To the right of the blocks and interconnects is a graphic of the PSoC IC for this project. This is the same graphic that appears in the project's data sheet. The I/O names that you've selected from the bottom left of the interconnect view. The pins are also color coded to reflect the chosen I/O settings.

Application Editor

The application editor is a multiple document text editor with advanced editor features. It is used to modify the code associated with your project. The application editor allows you to view an easy-to-navigate project explorer that organizes the project files into folders depicting their association with the project. The project explorer can be toggled on and off by selecting **Project** from the **View** menu.

There is also an output window pane which shows the results of the last build operation, debugging information and results of find in files operations. This pane can be toggled on and off by selecting **Output** from the **View** menu.

Cypress has done a good job with the editor functions. They've maintained standard shortcut keys to allow for quick editing of files. They've added a second find in files output window to allow you to keep the results of two separate searches for quick reference. A new file created from within PSoC Designer defaults to being added to your project. If this isn't desired, you can simply clear a check box to prevent it. They allow you, through editor options, to enable an output tab in your project view. This gives you quick access to the listing and mapping files.

The bottom toolbar gives you the ability to call up data sheets from any of their modules. Once the data sheet is opened, you can right-click on the data sheet and choose to open up the PDF file of the data sheet so as to leave it open for reference while you work in PSoC Designer.

There is an icon on the toolbars that looks like a wrench crossed with a hammer. This is the editor toolbox. You must be currently editing a file for this button to be active. It allows you to set the tab setting for your files. I don't see any equivalent menu option to change this setting.

Debugger

The in-circuit emulator (ICE) system connects to an emulator pod, which is a specially designed PSoC part that allows hooks to single step and set breakpoints within the part. The emulator pod has adapters, called feet, that allow adaptation to the different sizes of packages. This method was designed for through-hole systems.

The ICE system recently changed from a plain white unimaginative box to a new cube design that looks like it is a descendant of an iMac computer peripheral, complete with impressive blue LEDs. The connection to the emulator pods of the older ICE system is with an 8-pin RJ45 cable very similar to a standard network cable. This system can still be used with the new ICE cube via an adapter. However, the ICE cube also allows connections to 'Chip on Flex' technology. The newer emulator pods are of a surface mount design and are mounted on flex material. They are designed to be soldered directly onto your project board.

Since the feet system was a little fragile and tended to come apart, I shied away from using it. Some engineers actually glued the parts together to keep them from coming apart. The new flex system prevents that type of thing from happening. However, the difficulty of soldering the flex to your board and salvaging the flex for the next project can be fairly daunting also. I'm of a frugal nature and hate to risk damaging an expensive tool if I can help it. I usually resorted to leaving the emulator pod or flex system hanging in open air and work through other methods to simulate my circuit when debugging. There hasn't been a software simulator for the PSoC yet, although I've been told that there is one in process.

The debugger system of the PSoC has been useful in my experience, but also a little buggy. Setting a breakpoint doesn't guarantee you to stop at that point. You may stop a couple of instructions later. I've had the debugger seem to lose its connection with my file location and its spot in the code, leaving me to stare bewildered at the screen as the point of operation seems to jump around in the code with a mind of its own. I would expect it to recognize when something has changed and prompt me to fix it.

The events capabilities of the Cypress debugger are very impressive however. Events allow you set up logical conditions which when met will break operation, turn the trace on or off, or set an external trigger. This allows you to find difficult bugs in your code where a register may be getting corrupted or a stack overflow condition is occurring. It is a very powerful tool.

PSoC Designer is constantly being revamped for improvements, so annoyances are short-lived. However, it is important to note that you can't have multiple versions of PSoC Designer on your system at the same time. This means that projects done in older versions will need to be upgraded when moving onto a newer version. This has caused a couple of hiccups in my projects in the past. PSoC Designer might make changes to these files without giving you an option when you open older projects with newer versions of PSoC Designer so I would recommend making sure that older project versions are changed to read only so they aren't corrupted.

Different versions of PSoC Designer may load the files in different ways during the boot process. This has sometimes resulted in glitches on I/O pins during the boot process. You should take the effort to see what is happening during the boot process to make sure that this doesn't cause a problem with your project. Don't be too critical of how Cypress has done this, as it's very hard to make a procedure that works for every project that you may attempt.

Limitations of the PSoC

As with any part, the PSoC family of micros has certain limitations. The limitations of the PSoC's capabilities are often overemphasized, in my opinion. The PSoC family has a lot of capability despite its shortcomings. A little ingenuity and persistence will help you work around many of the limitations that you do find.

This chapter is not intended to discourage you from using the PSoC part, but will show you some of the limitations that you might encounter while doing your project. It's important to understand what the PSoC is capable of before embarking on your project, so that you don't work most of the way through your design and then run into a brick wall just before the finish line. Understanding what is possible in the PSoC will help you to decide what I/O you need to allocate and how much function you can squeeze into your next PSoC project.

Some of the limitations that I've listed in this chapter don't apply to all PSoC parts. Some of the newer PSoC parts have been redesigned to eliminate many of the more constricting limitations of the 25xxx/26xxx series. Therefore, as I discuss some of the limitations that I've encountered in my experience, I will include its relevancy to the different families and also include suggested ways to work around the limitation.

Analog Limitations

Analog Rail Limitations

The analog rail-to-rail limitation was the first troublesome limitation that I encountered. There is a limitation to what voltage range can be input and what voltage range to expect at the output as mentioned in the data sheet of the programmable gain amplifier. *The input and output voltage ranges of the amplifier do not extend to the power supplies, that is, they are not "rail-to-rail" op-amps. The linear output range extends to approximately 0.5 V from each power supply rail. The linear input range extends from approximately Vss + 0.5 V to V_{CC} – 0.8 V.*

This limitation applies to the 25xxx/26xxx series. The continuous time blocks are designed to work centered around AGND rather than towards either rail. The mistake that I suspect most will make isn't limiting the range of the input voltage, but remembering that the output voltage will fall under a similar limitation. Take for example a continuous time block that is configured as a programmable gain amplifier and you have the gain set at 2.00. In our example we have Vcc set at 5 volts. As your input voltage rises above 2.3 volts, the output is passing 4.6 volts. This exceeds the limitation of 0.8 volts from Vcc mentioned earlier. Therefore, you can't expect an accurate amplification in this instance after exceeding 2.25 volts at the input.

If you are using a continuous time block as part of a amplifier stage, and the amount of gain is causing the outputs to exceed the allowable output range, consider reducing the gain in that stage and make up for the lost gain in another way. For example, if the gain stage feeds an analog signal into another filter stage, perhaps you can adjust the gain of the switched cap blocks to compensate for the lower gain in your amplifier. If your amplifier is feeding an A2D converter, you might be able to increase the resolution of your A2D converter to compensate for the lower gain.

If you have an analog input signal that is close to the rail limitations, you should try to get that signal closer to AGND. I did some design on a motor controller project that required me to read the current going through the motor. Current would be measured by measuring the voltage across a 0.01 Ohm shunt resistor that was connected to ground on one end and the negative lead of the motor on the other end. The expected range of the motor current that I would need to interpret would be from 0 to 30 amps with 0.1 amp resolution.

You can quickly see that I would be dealing with small voltages on the current sense pin to interpret my current readings. I was able to add gain stages to help with the small voltages, but I was definitely going to fight the problem of having the input rail limitation with voltages that would be less than 0.5 volts. One method that Cypress suggested to me to get this signal into a more usable range involved a simple addition of two resistors as shown in Figure 4-1.

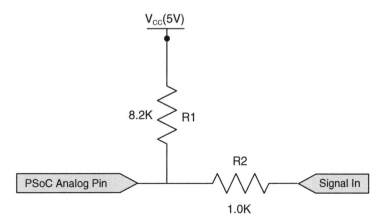

Figure 4-1: Voltage Divider Input

Using the voltage divider rule, I can determine what voltage I would expect to see at the PSoC analog pin. When Signal In is at 0 volts, I would expect to see around 0.543 volts at the analog input of the PSoC. If the Signal In voltage were at 3 volts, then I would expect to see around 3.217 volts at the PSoC analog pin. This is shown in Equation 4-1.

$$PSoC\_Input = (Vcc - Signal\_In) \cdot \frac{1.0K}{1.0K + 8.2K} + Signal\_In \qquad \text{Equation 4-1}$$

In order to derive the voltage change of the Signal In pin, I need to solve the equation above for the Signal In. I've substituted the known values into the equation above to make the solution simpler.

$$PSoC\_Input = (5 - Signal\_In) \cdot \frac{1.0K}{1.0K + 8.2K} + Signal\_In$$

$$PSoC\_Input = Signal\_In \cdot .891 + .543$$

$$Signal\_In = \frac{PSoC\_Input - .543}{.891}$$

Now that I have the equation that I'll be using in my project, I can insert the numbers that I've calculated from my circuit above.

$$\text{Signal\_In} = \frac{PSoC\_Input - .543}{.891}$$

$$\text{Signal\_In} = \frac{3.217 - .543}{.891}$$

A quick check on the calculator shows that I end up with the 3 volts that I expect.

This method adds some more calculations for me to go through to measure my current accurately, but it provides a way for me to use the suggested circuit from Cypress to measure my current and doesn't require more than a couple of resistors.

The newer versions of PSoC greatly improved this limitation. Those parts will have an input range of Vss to Vcc. The output swing of the continuous time blocks is from Vss + 0.05 to Vcc – 0.05 volts.

Analog Output Buffers

The analog output buffers allow an analog signal to be output directly to a pin on the processor. There are specific pins that are able to do this and only one pin per analog column. Therefore, you should check to make sure to reserve these pins first in your design if you plan to use an analog output. The analog output buffer pins also fall under an output limitation. This limitation is stated in both the 25xxx/26xxx and the newer 24xxx/27xxx parts' data sheets. They only guarantee the output to go from 0.5 * Vcc – 1.3 to 0.5 * Vcc + 1.3 volts. The actual parts that I've used have done much better than that and have gotten closer to the rails. However, I wouldn't plan on getting better than the data sheet if you are going for a high volume production setting. You should also note that reducing the load on the output of these buffers would help them to output voltages closer to Vcc and GND. I would recommend that you have some kind of external buffer if you plan on driving much current.

Another characteristic of the analog output buffers is that they will draw more current as they get closer to the positive and negative rails. This caused a problem for me when I was using the analog output buffer to send the state of a continuous analog block configured as a comparator out the analog buffer to an external pin on my project. The comparator is going to output either GND or Vcc.

Constantly trying to output one rail or the other at my analog buffer was enough in my project to disrupt the power supply and cause the project to misinterpret a small analog input signal. Fortunately, the output was a debugging option on my project and I was able to make a simple code change to shut off the output. If you need to get the output of a comparator to an external pin, allot appropriate digital sources in order to feed the comparator bus out a digital output rather than using the analog output buffers.

My design was a great example of a project that was already heading into production that was saved through the reconfigurability of the Cypress PSoC. A board change at that point would have been costly and time consuming.

High Input Offset Figure

The input offset is a variance away from the ideal balancing of inputs of an op-amp and is multiplied by the closed-loop gain of an op-amp. You could think of it as another voltage added to your input. It is multiplied by the gain of the amplifier and shows up on the output as a voltage that much larger. The input offset for the 25xxx/26xxx series of PSoC can get up to 30 mV although the typical offset should only be around 14 mV. With a gain of 1, this only equates to 1.5 counts on an 8-bit A/D sample and 25 counts on a 12-bit A/D sample. The newer PSoC parts have a much smaller input offset. The typical offset of the 27xxx part is only around 3.5 mV. The difficulty of having an input offset voltage becomes much more apparent however as you increase the gain, since the offset at the output increases proportionally with the gain of the amplifier.

One of my earlier projects with the PSoC involved conversion of an analog signal that was measured in small fractions of a mV. The 25xxx/26xxx parts were the only parts available at the time. In order to amplify the number to a usable range for the 12-bit A/D, I used two amplifier blocks to attain a higher gain than was available with one block alone. My design fed the signal through a programmable gain amplifier, then feeding the output of this amplifier through a second programmable gain amplifier as seen in Figure 4-2.

The signal that I was interpreting was referenced to ground, so we added a voltage divider circuit to our incoming signal to pull it away from the ground reference.

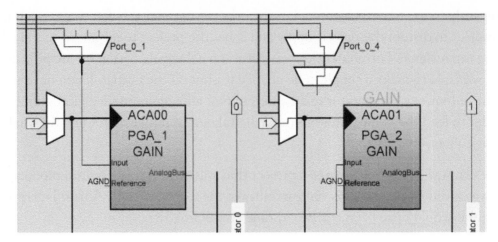

Figure 4-2: Daisy-Chained Gain Stages

Cypress suggested such a circuit to bring the analog signal to a level that wouldn't be bothered by the rail limitations of the 26223 part that we chose for the project. (Refer to the rail limitation section earlier in this chapter.)

Due to the small voltages and larger gains that I was applying to the input voltage, I needed to have some way to reduce or eliminate the input offset of the gain stages. I accomplished this by tracking an offset value that was measured at a time when I knew that there wouldn't be an input voltage at the pin other than the offset created by the voltage divider circuit. By storing the input offset number that was present at the output of the gain stages at that point, I was able to subtract that offset number from the readings of the analog-to-digital converter during operation. This allowed me to determine the reading of the analog signal without the added error of an unknown offset voltage.

Gain Deviation

My project to measure a small analog voltage mentioned in this section ran into another snag that had to be dealt with. This is the gain deviation of the continuous analog blocks. The gain deviation is affected by two factors: the amount of gain and the power applied to the block. This is best illustrated by a graph taken from the 26xxx data sheet for the programmable gain amplifier as seen in Figure 4-3.

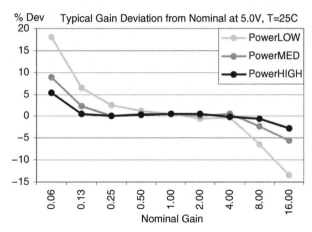

Figure 4-3: Gain Deviation

As you can see by the graph, you could experience a large variance in the actual gain value if you set up a large gain at low power. The easiest way to eliminate most of this error is to use high power to blocks that have a large gain. In fact, my advice is to set all analog blocks up as high power when you start your design. If a power issue arises after your design is working, then look to see where you can conserve power. Because the power setting of the block is simply a value in a register, you can easily set the block to a lower power setting or turn the block off during times of inactivity or in low-power modes when gain accuracy may not be as important.

The data sheet for the 26xxx part shows a variance from the gain that can be expected by mentioning gain variances for a few of the levels for high power only. It refers to the chart for a more complete picture. Please note that the gain tends to be lower than nominal at high gains and higher than nominal at gains less than 1.

The newer parts that use the improved analog show a great improvement with gain variation. Unfortunately, they have chosen not to include a graphical representation of how this gain can vary like they have in the 26xxx data sheet. They simply give typical values for various gains. The values from the 27xxx data sheet are as follows.

Gain Setting	Typical Deviation
48	4.0 %
24	2.2 %
16	1.2%
4	0.6%
1	0.3%

Cypress has definitely improved the variation of the gain. The deviation of gain at a gain of 16 is now expected to be just 1.2% where the data sheet of the 26xxx parts shows that you can expect an error of 5% at this setting. Note that the numbers listed in the table of the 26xxx parts show a grimmer outlook on gain deviation than the graph does. I don't know why there is a discrepancy, but I would design for the worst possible scenario.

Let's return to my project where I need to measure a very small signal. The variation in gain can definitely present a problem. If the error were to follow a more standard deviation bell curve type of error similar to a statistics problem, then the problem might be addressed by daisy chaining several small gains together. This would tend to even out the distribution and give a smaller overall error on average. However, the gain deviation tends to be in one direction. This would negate that effect. Looking at the percentages of gain error, I would even suppose that it might be slightly worse to take that approach.

In my project, gain error was a concern, so I looked at two possible solutions. The first is to simply use the expected deviation amount in my calculations. I believe that this would suffice for most projects. The more complex approach is to try to calculate the deviation amount. This is accomplished by taking a reading of a stable voltage at a gain of 1. After getting that reading at a gain of 1, change the gain the higher value that you expect to run at and then take a reading again. Compare that reading to the initial gain reading taking into account the differences in gain to calculate the expected gain error. There are some limitations in this of course, particularly if you need a lot of accuracy in your reading, but it will help you explore options to determine the true gain deviation of that part.

Switched Cap Blocks

The switched cap blocks are built around the concept that you can switch a capacitor in a way for it to look like resistor. Variations on how you switch the capacitor will let you change the resistive value. This allows a configurable analog system that can be very powerful, yet has limitations by nature.

Syncing Signals

Since you are switching a signal, the desired output isn't continuously present on the switched cap blocks. It's only present for a particular time. This means that you need to synchronize with the signal as you pass it from block to block. Fortunately, Cypress has made this extremely simple and handles most of the work for you. If your design requires adjusting on when it samples the signal, you only need to be worried about two times where the sampling will occur. This is denoted in PSoC Designer as *clock phase*. The two settings for clock phase are normal and swap. You will see this setting on modules such as the 8-bit delta-sigma analog-to-digital converter (ADC). I have been able to leave this setting at default (normal) for all of my projects with no problem.

Clock Limitations

Cypress has done an excellent job of trying to make these blocks as flexible as possible with limited resources and without making the blocks so complex that they become unusable. The clocking source for each column of analog is selectable. Variations of the clock allow you to set the centerline frequency for your filters. Since the clocking is necessary for the filtering action, you can't have true high pass filters. Signals can reach a level where they are too fast for the clock sources available and they will be distorted at the output. I haven't had to worry about this limitation in my projects so I'm not experienced on where those limitations fall. However, the filter design wizard directs me not to exceed 1 MHz in my filter design.

The selected clock for a column of analog blocks applies to all the blocks within that column. I first became aware of how this could cause problems by having a filter set up in the same column as a 6-bit successive approximation register (SAR) ADC. The filter required a slower clock to filter my desired band region, but I neglected to note that the SAR ADC was also in the same column. The SAR operates by stalling the processor to make its calculations on what level the analog signal is at. Since the amount of time that the processor stalls is directly proportional to the clock rate of the analog clock, I found that my chip would spend an inordinate amount of time whenever I tried to get a new sample. The sad twist of fate was that the sample was being taken from within a timed interrupt and the amount of time to process a sample exceeded the period of time before a new interrupt came pending. As you can imagine, the project didn't work very well, until I discovered the problem and made arrangements to resolve it.

Placement Limitations

Be forewarned that the placement options of multiple block modules are not infinite. This is due to both the architecture of the different switched cap blocks and to the arrangement of the switched cap blocks. The 25xxx/26xxx PSoC parts contain type A and type B switched cap blocks. The 27xxx and all other newer parts have type C and type D switched cap blocks. There are equal numbers of each type of blocks; the arrangement of the two types of blocks within the interconnect view is similar to the layout of a checkerboard with its black and red squares. This checkerboard arrangement allows the user to get the most variety in block arrangement within a particular part providing columns with the A type block on top and columns with it on the bottom. It also allows you to choose a column that has an AB arrangement when looking from left to right, or a BA arrangement. By trying to give you all possible arrangement types, you end up limiting the number of any given particular arrangement.

The limitations of block arrangement caused me a problem recently where I wanted to set up four band-pass filters directly through the chip and out the analog output buffers. I was not able to set up all my outputs as desired because of the layout of the blocks on the chip. On the columns that have the type C blocks on the bottom, I have to use the analog bus in order to carry the signal from the output of the continuous block in that section down to the input of the filter. My design, however, also needs the output of the filter to be sent on the analog bus to the output buffer. That leaves me in a quandary, as I can't use the bus for both tasks simultaneously. I haven't spent adequate time to find all possible solutions around this problem, so there might be an easy solution that I haven't found yet.

The placement limitations exist in both older and newer PSoC parts, however, they aren't exactly the same. The newer PSoC parts had improved flexibility on how the blocks can interconnect. I would suggest that any new designs use these newer parts for the added flexibility. The second important design rule that I've developed is not to commit to a project until I have placed the needed blocks within the project. In my example project of the preceding paragraph, I thought that placing four band-pass filters would present no problems. I only discovered my mistake after actually placing the blocks. Only then did the light go on, which left me returning to the design group with the disappointing news, which caused unnecessary redesign work.

It only takes a few minutes to place the blocks that you think you will need for a project. I would suggest that you take those few minutes immediately after you have enough information to know which blocks you will need to place. This way you will not be surprised towards the end of the game by a roadblock that inconveniences everyone.

Signal Routing

One improvement to the newer PSoC parts that I believe was well needed is the ability to route an analog signal directly into a switched cap block. Some possibilities can be seen by trying all possible placements of an 8-bit delta-sigma ADC block in a 27xxx part. You can see that there is a new input option that appears on the leftmost two switched cap blocks and on the rightmost two switched cap blocks. The blocks on the left allow you to choose port2[1] and the blocks on the right allow you to choose port2[2]. I imagine that the inputs are available with blocks on the extremities because these blocks have less interconnects routed through them than the center blocks do. Bits 1 and 2 of port 2 are used to connect to the A mux of the switched cap block. This is why they appear for the ADC. If I were to place a band pass filter in its possible placements that use the C mux, then I would see the possibilities of using port2[3] to input to the top left switched cap block or port2[0] for the bottom right switched cap block. If you don't need added amplification of an analog signal, this becomes a slick way to process a signal without tying up another continuous time block.

Noise in Analog Area

Some critics of the PSoC analog system have talked about how coupling of switching digital signals to the analog sections of the PSoC make the use of the PSoC quite limited. I must say that I haven't seen a problem with my design work with the PSoC micros. I have seen some of the noise come through, but I've only seen it when I haven't had things set up correctly.

Digital Limitations

The PSoC family has two types of digital blocks. The 25xxx/26xxx parts use the DBAxx and DCAxx blocks. The DCAxx blocks are used for communication, in addition to being able to perform normal timing, PWM and other functions. The 24xxx/27xxx/29xxx and other newer parts use the DBBxx and DCBxx blocks. These blocks have some enhancements. In addition, Cypress has enhanced the interconnect-ability around the digital blocks, which allows you to achieve more complex projects.

The best way to prevent panic in your design is to set out all the blocks needed for a project and connect the blocks before committing to a design. It can be extremely difficult, especially when first starting to design with PsoC, to foresee all the scenarios that are possible without a snag.

When Cypress created PSoC Designer 4.x, they did a great job into making the interconnections of the blocks visible to the user. The older Designer versions were a bit more cryptic on exactly what was going on. You can learn a lot by simply clicking around in the interconnect view of Device Editor to explore how things are connected.

Communication Modules Had to Exist on the Higher Nibble of the Digital Blocks.

If you want to use the digital blocks for communication, you will need to place the blocks in the DCAxx communication blocks on the 25xxx/26xxx parts or in the DCBxx blocks for the 27xxx or newer parts. These communication blocks comprise half of the digital blocks available on most parts. Therefore, if you are going to use a CY826443 part, you will only be able to use four blocks for communications function. This allows you to have two full UARTs. Likewise with a 27443 part, you are only able to have two full UARTs. Now if you want to only use the receive portion of a UART, then you don't have to use two full blocks for the UART function. You can insert an 8-bit serial receiver block and simply use the one block to receive, thereby freeing up an extra digital block in your project.

Input/Output Pins for Communication Blocks

On the 25xxx/26xxx series of parts, the communication blocks were placed in the second row of logic as viewed in the interconnect view of the Device Editor. This means that these blocks can only be connected to the global buses 4–7. Therefore, you are only able to connect these pins out the high nibbles of any particular port.

When you are first setting up a design that needs to use communication blocks, make sure to reserve pins in the high nibble of your ports for the communication function. It's also imperative to make sure that you use unique bits within the nibble for particular functions. For example, you can't have two different transmitters outputting to port1[7] and port2[7]. Only one of the transmitters may connect to the global bus bit 7, not both. You will need to move one of the transmitters to a different bit, such as port1[6], to be able to route your signal to the output pin.

You can, however, use the same bit for one input and one output. For example, in one project you are able to place a UART and define port1[7] to be transmit and port2[7] to be receive. This is because the global bus bit 7 input and global bus bit 7 output are two different lines and therefore you won't have a conflict.

When Cypress created the 27xxx series, and with other newer parts, they rearranged how the blocks are organized. Rather than have one row of normal timing blocks and one row of digital communication blocks, they interspersed the communication blocks. You have two timing blocks and two communication blocks on each row of digital blocks. They've also revamped the interconnections between the blocks and the pins to allow you to put the communication signals out to any pin on the processor.

Timing Inputs

Digital input blocks are able to access timing from 16 different sources. The need to plan out timing can come early on in the game. You should lay out your blocks and resources as quickly as possible to make sure that you have everything covered. Some of the resources that you think may be there could end up not being there. For example, if you want to implement a UART, you will also need to provide a clock input to the UART that will function as the baud rate generator. That means that the clock input for the UART needs to be eight times faster than the desired bit rate of the UART. This was a surprise for me on my first project that implemented a UART, because it meant digital block usage that I wasn't expecting. It was actually the UARTs that helped me become more familiar with what was possible with the timing inputs of the PSoC.

Several clock sources are available to the digital blocks as listed in the global resources. In addition to the global clock sources you can also use the output signals of other blocks as the clock source. This could easily be needed, especially if you want to get to a lower baud rate. This can present a bit of a dilemma if you are trying to run two UARTs where the UART baud rates require clocks that can't be supplied by the 48 MHz, 32 kHz, or one of the 24Vx numbers. V3 is available on the newer parts, but I use that resource for a general-purpose timing in my projects to define other elements of my projects.

Since the UART modules require two digital blocks for each instance, you will have consumed all four digital communication blocks. This doesn't leave you any blocks left in the high nibble for timing purposes. The DBA03 has an interconnect available for you to use it as a timing block resource across all digital blocks on the 25xxx/26xxx parts. It can be used as the clock source for both UARTs if both your UARTs have the same baud rate. If they don't have the same baud rate, then you will need to implement global output 0. It is the one other digital block clock source that jumps the gap, going from the lower nibble to the higher nibble. (Note: You can also use global output 4 to get signals from the higher nibble to the lower nibble.) In this situation then, you have one timer in a lower nibble digital block that is outputting a needed clock frequency for one UART and one timer in DBA03 outputting the frequency needed for the other UART. That only leaves two blocks left if you implement the timing in an 8-bit timer. It also means that if you want to use a digital block to get a signal to the outside world, you shouldn't use bit 0 of any port to do it, because global out 0 is being used for an internal signal.

This situation is a good example of how the resources of the PSoC micro can easily be expended. I work with ZiLOG® Z8 Encore! processors also. They have two UARTs with their own baud rate generators, in addition to several timers that can function as PWM generators. The ZiLOG Z8 Encore! also includes other communication features such as SPI and I²C. It would be impossible for me to use the 26xxx or 27xxx family of

PSoC parts and duplicate all the features of the Zilog Z8 Encore! parts in one configuration. I don't have enough resources to do it. (Note: the 29xxx parts have 12 digital blocks and get you closer to this goal.) This might come into play as you plan out your project. However, the Zilog Z8 Encore! can't be configured to do everything that the PSoC can do either. This situation is yet another good example of why you should lay out all your blocks and interconnects as quickly as possible when planning a project.

The final, albeit crude option for routing timing signals is to have the desired timing signal output to an I/O pin as a global signal and then bring the signal back in another I/O pin an a global signal into the target digital block. I feel that this is a solution of last resort as it is a costly use of global resources and I/O pins, but sometimes it is the only solution left and may not be that big of a deal on your project.

The newer 27xxx series and other newer PSoC parts change the structure of the digital blocks by putting communication blocks in both rows of digital blocks. This gives you a couple of other connection options because you don't have four communication blocks in a row. You can place timers in the blocks immediately to the left of the communication blocks and use the previous block option for both UARTs. It also gives you the opportunity to use two 16-bit timers to time your UART blocks if desired.

Interconnects on the Newer Parts

The interconnects between digital blocks and I/O pins or other blocks were redesigned in the newer PSoC parts. The older 25xxx/26xxx blocks were set up to connect to global buses of the lower nibble if the blocks reside in the first row of logic or use the global buses of the higher nibble if in the second row of logic. For example, if you place a timer into DBA01, a block in the first row of logic, it will have among its available clock options: GLOBAL_IN_0, GLOBAL_IN_1, GLOBAL_IN_2, GLOBAL_IN_3, GLOBAL_OUT_0, GLOBAL_OUT_1, GLOBAL_OUT_2, and GLOBAL_OUT_3. There are also the global clocks available as options, such as 48 MHz, 24V1, 24V2, and 32 kHz. Then there is a broadcast line that originates from DBA03. The DBA03 broadcast and these other options are available to all digital blocks. The remaining options depend on what block you are in. There is the interconnect that connects to the output of the block immediately to its left (as viewed in PSoC Designer). It will be referred to as the *previous digital PSoC block* clock select in the data sheet. These options are described graphically in Figure 4-4.

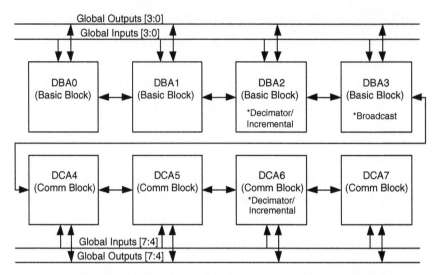

Figure 4-4: Digital Block Arrangement of 25xxx/26xxx Parts

The final interconnect depends on what row of logic you are in and is not graphically shown in Figure 4-4. It is GLOBAL_OUTPUT_4 if you are in blocks DBA0–DBA3 or GLOBAL_OUTPUT_0 if you are in blocks DBA4–DBA7. This allows another path for you to get a timing signal from one row of logic to another as was shown in the previous section.

By using the newer parts, your options have changed somewhat. I think the best way to illustrate this would be to show the clocking options of an 8-bit timer placed in block DBA01 on a 26xxx part and an 8-bit timer placed in block DBB01 on a 27xxx part.

26xxx	27xxx
DBA03 Broadcast	Row_0_Broadcast
DBA00	DBB00
48MHz	SysClk*2
24V1	VC1
24V2	VC2
CPU_32KHz	VC3
GLOBAL_OUT_0	CPU_32kHz
GLOBAL_OUT_1	Row_Output_0
GLOBAL_OUT_2	Row_Ouput_1
GLOBAL_OUT_3	Row_Output_2
GLOBAL_OUT_4	Row_Output_3
GLOBAL_IN_0	Row_Input_0
GLOBAL_IN_1	Row_Input_1

26xxx	27xxx
GLOBAL_IN_2	Row_Input_2
GLOBAL_IN_3	Row_Input_3
Disable	Disable

The very first clock option is different on the newer parts. It lists a row broadcast rather than a broadcast from a particular block. This option Row_0_Broadcast would change to Row_1_Broadcast if the timer were placed in the second row of digital blocks. These row broadcast options are more flexible than the DBA03 broadcast. Clicking on the Row_0_Broadcast will allow you to choose the source of the broadcast signal. You can choose the output of any of the four digital blocks on that row, or you can choose the Row_1_Broadcast as the source signal. Likewise, the source signal for Row_1_Broadcast can come from any of the four digital blocks on the second row of digital blocks or it can come from the Row_0_Broadcast. This system will allow you to get signals transferred from either row of digital blocks to the other regardless of which block you want the signal source to come from. In addition, you can use the row broadcast signals independently from each other to allow you to interconnect signals from nonconsecutive blocks on both rows without wasting the global connections.

The next option for the previous block is identical. The SysClk*2 source does give a little added flexibility as the SysClk source does not have to come from the internal oscillator. Please see Appendix A – *Global Resources* to learn about SysClk.

The row input and row output interconnections provide more flexibility than the previous global buses. Both rows of digital blocks are now able to go to both nibbles of all output pins. The rows are subdivided as being Row_Input_0_0 on the top row of the first row of digital blocks and Row_Input_1_0 on the top row of the bottom row of digital blocks. When selecting the parameters for your digital block in the user module parameters section of Device Editor, you will only see the resources described as Row_Input_0 as the number of the digital block the module is placed in will determine which row of interconnects it is using.

Row_Input_0 has the ability to connect to either bit 0 or 4 of any port. Row_Input_1 has the ability to connect to bit 1 or bit 5 of any port and so on through Row_Input_4. The two columns of global pins it is connecting to have the titles GIO and GIE. These are acronyms for global input odd and global input even. Port 0 is an even port,

Port 1 is an odd port and so on. More details on how to use these interconnects and other new features are discussed in Chapter 5 – *Improvements of the PSoC* and Chapter 7 – *Interconnects*.

Improvements of the PSoC

Cypress Semiconductor Corporation is aware of the shortcomings of their product and have been working hard to produce newer generations of parts that overcome these shortcomings and offer new families of PSoC that have varied combinations of digital and analog capabilities to meet the needs of a larger microcontroller market. They have been able to increase functionality and still keep the price of their parts competitive with traditional microcontrollers. This chapter will discuss in a little more detail some of the improvements in the newer families of Cypress Semiconductor Corporation products. There are yet some more parts that are in the design stage that will be available by the time this book is printed. I will not, however, include those parts in this edition.

Analog Improvements

Cypress Semiconductor Corporation has endured a lot of scoffing at the analog limitations of their analog blocks in the first version of PSoC microcontrollers. I don't believe that they were misleading at all when they introduced the parts to me. I was told that the PSoC microcontrollers would be able to perform a lot of different functions; but the analog capabilities of the part had limits and they would be the first to admit if the design would be better suited to an external analog component with more capabilities.

They held true to their word. I think that I did accomplish more than I first thought could be done by some inventive methods that Cypress suggested, but to work around the limitations did take some additional time and sometimes required a bit of a compromise in order to make them work. Cypress Semiconductor Corporation knows that real world engineers cringe at words like *compromise* and *additional time* so they have worked feverishly on improving their analog subsystems. They have done an excellent job with their second generation of parts. I look forward to even newer designs on the horizon which will add function and capability beyond what we have now. The one surety is that Cypress Semiconductor Corporation is moving forward and doesn't show any signs of stopping.

PSoC Core

As the first Cypress PSoC micros hit the marketplace and started to be used the little problems that always accompany a new design started to crop up. Cypress kept up numerous errata and application notes to document and provide work arounds for the various situations. They were able to correct many of these problems on their newer parts and have improved the capabilities of the part so that you don't have to be as concerned about a potential problem with the newer families of PSoC.

24 MHz Alignment Shift Eliminated

Higher operational speeds require tighter specifications on how the microcontroller is constructed. Since all circuits have some component of resistance, capacitance, and inductance, you need to ensure that the circuit is built around tight enough tolerances to be certain that the internal signals of the microcontroller will reach an adequate voltage level in the time needed in order to work consistently. When Cypress produced the first families of parts (25xxx/26xxx), they noticed that at the highest speed possible for the instruction clock (24 MHz), you could have a problem with switching pages of Flash fast enough to guarantee the appropriate read of Flash if an instruction were to cross the page boundary.

They first came out with an application note describing the danger and how to look for a work around. As you can imagine, however, trying to decipher the situation where an instruction straddles a page boundary is quite daunting. It was a painful circumstance to tell everyone that you would need to stay at the lower 12 MHz operating speed. Cypress therefore worked on incorporating a check for the potential problem into the next version of PSoC Designer. A simple checkbox would tell PSoC Designer to warn the user when this type of problem could exist. Another checkbox would instruct PSoC Designer to alter how the files were linked together to fix your project so that this situation would not exist. This method worked for most situations. The only time that I had difficulty is when a project required use of all but a few bytes of Flash in the microcontroller. That situation would leave PSoC Designer with very little maneuvering room, to the point that it could not find a solution. Fortunately, I didn't need to run those projects at 24 MHz, which made the situation disappear.

The newer parts were redesigned in this area to make the 24 MHz alignment shift unnecessary. The part can run at 24 MHz in all situations.

System Clock

Previously the internal main oscillator (IMO) was generated by an internal 24 MHz oscillator only. The newer parts don't change that, but they give you an option to provide an external clock signal to be used for the internal main oscillator. The external clock signal shouldn't go any faster than what is suggested when using the internally generated oscillator value; in other words, 24 MHz for 5 volt operation and 12 MHz for 3 volt operation. However, you can go anywhere below that.

Using this method, if a more accurate clock is available, then you can use that clock source instead of relying on the internal oscillator. You also won't have to deal with the time delay of phase locking to the external crystal and can use one pin for a clock source rather than the two pins to drive the crystal.

You also can feed in some more desired frequencies. For anyone who has done serial RS232 communications, you know that trying to get a baud rate of 300, 1200, 9600, etc. isn't an easy integer divide with 24 MHz. You will have a slight error in your actual baud rate. This is acceptable in many situations, but in the situations that need to have a more accurate system, you buy crystals that work at a frequency that is a multiple of your desired baud rate. Now you have the opportunity to input that frequency in the PSoC microcontroller so that the entire system will be able to generate several of these baud rates with complete accuracy.

VC3

VC3 is a new timing tap available on the newer PSoC parts. It works in concert with VC1 and VC2, which are very similar to 24V1 and 24V2 of the older PSoC parts. It differs from VC1 and VC2 in that it has a larger range and can be used as an interrupt source.

The addition of VC3 was most welcome indeed for my projects, as I was always using at least one digital block to house an 8-bit timer for a general-purpose timed interrupt. VC3 is now able to fill that need, giving the potential to divide the source clock from 1 to 256 times. There are certainly other functions that VC3 can serve. First of all VC3 has four different potential clock sources: SysClk, VC1, VC2, or SysClk*2. This gives it the chance to output an additional signal of a different frequency in the same range area as VC1 and VC2, or you can get a signal that is much lower than these two by having VC2 as a source. By using all three VC clocks together, you can produce a frequency of less than 400 Hz with the internal oscillator and without using any digital blocks.

VC3 also will provide an interrupt at its terminal count. Since I like to have projects that have events happen in a timely manner at regular intervals, I use the VC3 clock as a interrupt source for my fastest desired interval and then divide that interval down further with software. This works very well and saves the use of the digital block that I used in the past for this purpose. Appendix B – *Project Walkthrough* illustrates an example project of this type.

Digital Blocks

The first dramatic improvement to the digital blocks is the placement of the basic digital blocks in relation to the digital communication blocks. The arrangement of the 25xxx/26xxx parts had all the basic digital blocks on one row and all the digital communication blocks on the other row.

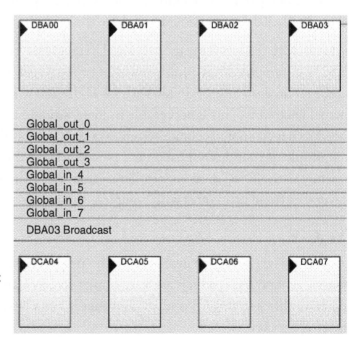

Figure 5-1: Arrangement of Digital Blocks on 25xxx/26xxx Parts

The newer parts changed this arrangement to a more desirable layout by having half the digital communication blocks on either row.

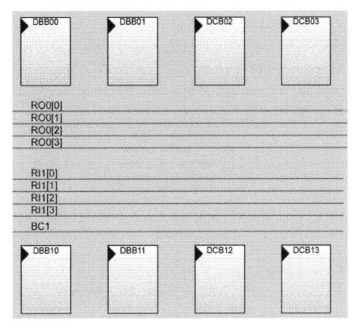

Figure 5-2: Arrangement of Digital Blocks on 27xxx Parts

Because there are interconnects in between the blocks that allow you to choose the output of the previous block without using the digital array of interconnects, there is some more flexibility to having the digital communication blocks immediately to the right of basic digital blocks. The last chapter mentioned briefly that this would allow you to implement two full UARTs in the PsoC, have their clocking signals be generated by the digital blocks immediately to their left, and not use the digital array signals in the process. This was not possible on the older parts without using a global bus line or the DBA03 broadcast signal.

Another improvement with the digital blocks in the new PSoC families is the addition of the synchronization parameter. Because you are now able to enjoy more clocking options with your digital blocks, it's important to have the blocks synchronized correctly to control skew and make sure that the process of digital signals from block to block will meet the correct timing requirements. The data sheet for the older PSoC parts mentions synchronization to the internal 24 MHz clock which would suffice for most clocking options, but would not work as well if you chose 48 MHz for the digital clock.

The new PSoC families allow you to input an external signal for SysClk, so synchronization with the internal 24 MHz clock might not have any practical effect. Therefore the option is given to you to synchronize to SysClk so that you will be synchronized correctly regardless of the source of SysClk. Also you are given the option to synchronize to SysClk*2 if you are going to use the 48 MHz option generated from the internal 24 MHz oscillator or the corresponding frequency from your external clock source.

Each allowed synchronization follows simple guidelines on when to select that particular option.

Use SysClk Direct – Use this option if the clock source of your digital block is SysClk.

Sync to SysClk – Use this option if the clock source is derived from SysClk, for example, VC1, VC2, or VC3 when the source of the VC3 is SysClk. This also applies if the clock source for your block is the output of another digital block that uses a clock derived from the SysClk.

*Sync to SysClk*2* – Use this option if the clock source is derived from SysClk*2, for example, VC3 with the source of VC3 is SysClk*2, or if the input clock is SysClk*2. This also applies if the clock source for your block is the output of another digital block that uses a clock derived from SysClk*2.

Unsynchronized – This option isn't selected in most designs, but may be desired if the clocking signal for your digital block is derived from external circuitry that doesn't originate within the PSoC. Synchronization is then left up to the user.

The data sheet also gives a few warnings on configurations that you are able to load into your part that you should avoid. Don't set VC1 and VC2 both to the value of 1 and then try to use VC2 as a clock source for your digital block. This is the same frequency as SysClk and you should use SysClk. Likewise, don't select a value of 1 for VC3 and use VC3 as a clock source. Use SysClk or SysClk*2 as the clock source for your desired frequency.

Improved Interconnects

Improvements to interconnects between digital blocks were welcome improvements to the PSoC families. Previously, there were 16 global bus lines. Eight lines are used to input signals and eight lines are used to output signals. It was possible to bring a signal in on one global bus line and connect it to multiple blocks. You could also use a global output line that would pass the output of one digital block to an external pin and even use this output bus as the input for another digital block. Each global bus would connect to the corresponding bit of any particular port. For example, if I want an input signal from an I/O pin to come in on Global Bus In 4, I would need to connect that signal to port 0 bit 4, port 1 bit 4, or port 2 bit 4, and so on. This allowed a lot of flexibility, but as you started to get more and more complex systems running in the PSoC, you found that you would be running out of pins or you couldn't pipe the signals to the blocks as desired without changing the hardware around the pins to make everything optimized. Changing hardware is exactly the type of thing that you want to avoid, so Cypress found some ways to improve these interconnects.

The first thing to notice about the interconnect array of the newer PSoC parts is that the 16 lines surrounding the digital blocks are not connected by default to any of the global buses. There is an advantage to this in that you are able to interconnect two digital blocks and still leave all the global buses free for other uses. Connecting between the global bus array is now through an input multiplexer or an output de-multiplexer. The input multiplexers are attached to the red input rows that are positioned above the digital blocks in the interconnect view. Clicking on one of these input rows will show the possible connections of the interconnect line to global input signals to the left of the digital blocks. The output de-multiplexers are connected to the blue output rows that are positioned below the digital blocks in the interconnect view. Likewise, clicking on the output row will display the possible connections to the global output signals.

This method allows a particular row to connect with twice as many locations as the old parts. In addition, the newer interconnects allow you to have two global input signals on the same bit from two different ports. Please refer to Chapter 7 – *Interconnects* to learn more about interconnects.

In addition to the interconnects, the DBA03 broadcast of the older parts has been improved to be a row broadcast signal that allows any of the digital blocks on that row to be broadcast to other blocks on that row. There is also a separate broadcast signal on the bottom row of digital blocks that acts in the same way. The two broadcast signals can be linked together to allow the broadcast signal from one row to be used by digital blocks on the other row.

LUTs in Key Locations

Digital outputs and comparator buses have logic look up tables that allow you to add some simple logic to their signals, which would previously have required a digital block. This was rather difficult on the older parts. I had a project that required the output of a UART module to be inverted. Since I didn't want to add the cost of external logic, I had to tie up extra global bus in order to pass the signal from one block to another block which would act as the inverter and then send the signal of the inverter out to the external circuitry. In addition, the extra digital block had to reside on the same row and thereby cost me a digital communication block rather than a digital basic block. There were some other possibilities that required me to send the signal out on an I/O pin and then back in another I/O pin, which wasn't very desirable.

Input and Ouput Pins

Chapter 2 – *Structure of the PSoC* mentions the addition of the DM2 register for I/O pins. This extra register is used with some more complex drive logic to give the additional open drain modes (including active low or active high states), slow strong drive, or high impedance analog.

The open drain modes are useful for implementing communication buses such as I²C or multiple master situations where you want to prevent damage to components without the addition of external circuitry to add the open drain method. You will still need to add the external resistor to assert the inactive state.

The slow strong drive removes the high frequencies generated by faster changing signals. This becomes very useful especially where the signal line is long and/or is transferring a large amount of current. These two situations tend to make an effective antenna that could be disruptive to other parts of your design or nearby electronics.

The high impendence analog (High-Z Analog) option will disconnect the digital input signal from the I/O pin circuitry. This reduces the current draw in your project. The data sheet refers to it as the zero power state. This is the default state at reset. It is important to note that you need to change this state to High-Z if you need to use the I/O pin as a digital input. The current data sheet does not show the actual input impedance improvement from between the High-Z and High-Z analog settings.

Continuous Time Analog Blocks

The continuous time analog blocks have been improved in their structure. Probably the most notable improvement is that the operational amplifiers that are used in both the continuous time blocks and the switched cap blocks are able to get closer to the rails with their signals. This was a sometimes difficult limitation of the older family of parts. The input voltage swing is listed to be from Vss to Vdd. The output swing is listed to get from Vss + 0.05 volts to Vdd − 0.05 volts. I have taken this number from the PGA data sheet. The general data sheet for the parts lists different numbers. I suggest that you adhere to the numbers that accompany the user modules. I believe that the numbers in these situations take into account other factors that may affect the capabilities of the components used by the module. I've also noticed that the modules and their data sheets have been updated beyond the available data sheet revisions for the PSoC parts.

Chapter 2 – *Structure of the PSoC* also mentioned two other important improvements: the input offset and the gain deviation. The newer PSoC parts have tightened tolerances to make significant improvements to keep the gain closer to the designed value by as much as four times as accurate. In addition, the PGA data sheet shows the input offset figure being 1/3 the previous value. Since the input offset affects the output voltage by the offset value multiplied by the gain, this is of great benefit.

In addition to the increased voltage ranges, the continuous time blocks allow for two more gain settings. These gain settings will allow you to amplify your incoming signal by 24 or 48. All other gain settings that were available on the 25xxx/26xxx parts are still available on these parts. Since bits 7:4 of the control register 0 are used to set the gain, all available bits in that control register were in use and could not be used to select the additional gain settings. In order to maintain backwards compatibility,

Cypress has made the additional gain settings by using an extra bit from control register 3. This bit is referred to as EXTGAIN.

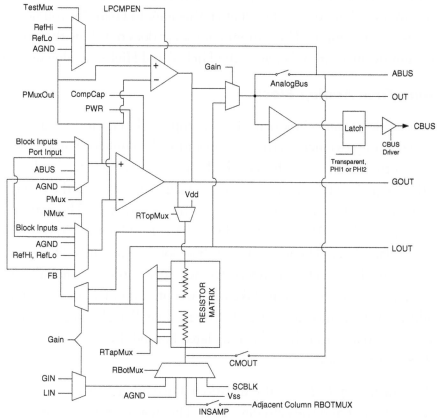

Figure 5-3: Continuous Time Block Type B

The addition of control register 3 allows control of some other added features to the continuous time analog blocks. The added features are a low-power comparator and resources to build an improved instrumentation amplifier.

The low-power comparator is at the top of Figure 5-3 and is enabled by the LPC-MPEN (low-power comparator enable) bit. This low-power comparator allows the user to select a basic comparator function when power consumption is more important than speed or offset. Care should be taken to power down the main op-amp when using the low-power comparator. This is done by setting the control register 2 bits 1 and 0 both to be 0. The power of the low-power comparator is unaffected by this shutdown, but you could force contention among the op amps if they are both running at the same time, as their outputs are tied together.

The added connections of RBotMux to the output via CMOUT and the added input of the adjacent column RBotMux via INSAMP allow for a new instrumentation amp topology as seen in Figure 5-4. The involvement of the CMOUT and INSAMP bits are shown in Figure 5-5.

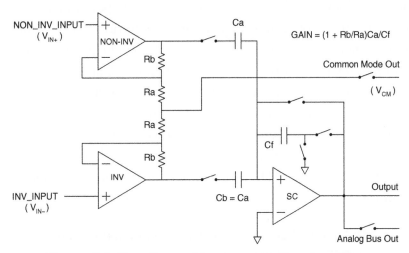

Figure 5-4: New Type of Instrumentation Amplifier

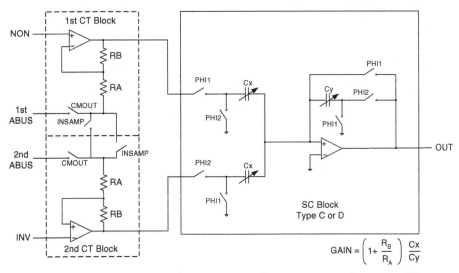

Figure 5-5: Structure of New Type of Instrumentation Amplifier

Previously, the instrumentation amplifier module only involved two continuous time analog blocks with the PSoC. This type of module is still available on the newer PSoC parts. The new topology, however, boasts an improved common mode rejection and can provide a higher gain for the output. In addition, the CMOUT connection allows the common mode to be output on an analog bus if desired.

Switched Cap Blocks

The switched cap blocks also benefit from the improved operational amplifiers and gain the same advantages of accuracy and range. In addition to these improvements, there are some added connection abilities. Some of the added connections are seen in the control register 1 bits of the switched cap blocks. Chapter 2 – *Structure of the PSoC* showed some of the similarities of the type A blocks of the 25xxx/26xxx parts and the type C blocks of the newer PSoC parts. You can see from the tables below that Cypress has utilized the reserved options of this register to allow more combinations of connections. These added connections are used to build better filters in a larger variety of configurations. The larger variety of configurations allows you to fit more function into a single PSoC.

Table 5-1 Switched Cap Type A Control Register 1 Bits 7–5

	ASA10		ASA21		ASA12		ASA23	
	A Inputs	C Inputs	A Inputs	C Inputs	A Inputs	C Inputs	A Inputs	C Inputs
000b	ACA00	ACA00	ASB11	ASB11	ACA02	ACA02	ASB22	ASB13
001b	ASB11	ACA00	ASB20	ASB11	ASB13	ACA02	ASB22	ASB13
010b	REFHI	ACA00	REFHI	ASB11	REFHI	ACA02	REFHI	ASB13
011b	ASB20	ACA00	Vtemp	ASB11	ASB22	ACA02	ABUS3	ASB13
100b	ACA01	Reserved	ASA10	Reserved	ACA03	Reserved	ASA12	Reserved
101b	Reserved	Reserved	Reserved	Reserved	Reserved	Reserved	Reserved	Reserved
110b	Reserved	Reserved	Reserved	Reserved	Reserved	Reserved	Reserved	Reserved
111b	Reserved	Reserved	Reserved	Reserved	Reserved	Reserved	Reserved	Reserved

Table 5-2 Switched Cap Type C Control Register 1 Bits 7-5

	ASC10		ASC21		ASC12		ASC23	
	A Inputs	**C Inputs**	**A Inputs**	**C Inputs**	**A Inputs**	**C Inputs**	**A Inputs**	**C Inputs**
000b	ACB00	ACB00	ASD11	ASD11	ACB02	ACB02	ASD13	ASD13
001b	ASD11	ACB00	ASD20	ASD11	ASD13	ACB02	ASD22	ASD13
010b	REFHI	ACB00	REFHI	ASD11	REFHI	ACB02	REFHI	ASD13
011b	ASD20	ACB00	Vtemp	ASD11	ASD22	ACB02	ABUS3	ASD13
100b	ACB01	ASD20	ASC10	ASD11	ACB03	ASD22	ASC12	ASD13
101b	ACB00	ASD20	ASD20	ASD11	ACB02	ASD22	ASD22	ASD13
110b	ASD11	ASD20	ABUS1	ASD11	ASD13	ASD22	ABUS3	ASD13
111b	P2[1]	ASD20	ASD22	ASD11	ASD11	ASD22	P2[2]	ASD13

The type B and type D blocks also share some similarities across the two types of parts. You will notice that both types have the same connections available. However, the tables from the data sheet list the blocks in a different order. These tables are recreated here in Tables 5-3 and 5-4 showing the difference in listed order, but on close examination, you can see that the connections are indeed the same.

Table 5-3 Switched Cap Type B Control Register 1 Bits 7–5

	ASB11	**ASB13**	**ASB20**	**ASB22**
000b	ACA01	ACA03	ASA10	ASA12
001b	ASA12	P2.2	P2.1	ASA21
010b	ASA10	ASA12	ASA21	ASA23
011b	ASA21	ASA23	ABUS0	ABUS2
100b	REFHI	REFHI	REFHI	REFHI
101b	ACA00	ACA02	ASB11	ASB13
110b	Reserved	Reserved	Reserved	Reserved
111b	Reserved	Reserved	Reserved	Reserved

Table 5-4 Switched Cap Type D Control Register 1 Bits 7–5

	ASD20	**ASD11**	**ASD22**	**ASD13**
000b	ASC10	ACB01	ASC12	ACB03
001b	P2[1]	ASC12	ASC21	P2[2]
010b	ASC21	ASC10	ASC23	ASC12
011b	ABUS0	ASC21	ABUS2	ASC23
100b	REFHI	REFHI	REFHI	REFHI
101b	ASD11	ACB00	ASD13	ACB02
110b	Reserved	Reserved	Reserved	Reserved
111b	Reserved	Reserved	Reserved	Reserved

PSoC Modules

Each PSoC module has its own data sheet, but I would like to mention some of the highlights of each module to give you a quick reference on what each module can do, placement considerations and other pertinent information that may not be immediately apparent. The modules are divided into the groups designed by Cypress. The modules listed were the modules currently available in PSoC Designer version 4.1.

Please note that this chapter is not intended to be a comprehensive source for module operation. Each module has its own data sheet that should be consulted when a module is placed in a project. After placing a module in a project, please be aware that the module will not work properly without executing some code. Each module has to have its start routine called to work properly. The start routine will set the enable bit on digital blocks and set the power of the analog blocks. Some modules, such as analog blocks, require arguments in the start routine in order to make them operate properly. Some digital block modules don't require any arguments to be passed to the routine. You just need to call the start routine. There are some modules that have empty start routines, but the routines exist anyway. The idea is to set a convention of calling the start routine for all modules before operation.

All API functions for blocks placed in your project will be generated by PSoC Designer. You have on option on the newest version (4.2) of PSoC Designer to have PSoC Designer automatically remove any API functions that aren't called by your project, in order to save space. Please be aware that it is a Cypress convention that the A and X registers may be modified by their API calls. It is also the method used to pass in and return parameters for these calls. If these registers need to be preserved, then you should take care to push and pop the registers around the API calls.

Analog-to-Digital Converters (ADCs)

The analog-to-digital converters of the PSoC are designed to reside in the switched cap blocks. It's a good idea to experiment with placement of the ADCs before the schematic for your project is finished, to ensure placement of the blocks in the desired location. The ADCs are designed to measure a signal centered around AGND and between the high and low reference points.

If you are expecting an analog signal to range from Vss to Vcc, then you need to make sure that the global parameter for ref mux is set to (Vcc/2 ± Vcc/2) or (Vdd/2 ± Vdd/2). If your ref mux is set to the default of ± bandgap then you may not get the answer back that you were expecting.

Note: Different files for PSoC Designer have listed the supply voltage as Vcc and others as Vdd. They are the same.

ADCINC12

The incremental ADC works by configuring a switched cap block as an integrator. The input voltage and the reference voltage are switched alternately into the integrator and a counter is set up at the output to count how many times the integrator is high. That count is processed through an interrupt at regular intervals. The upper bits of the result are handled in software, as this ADC only uses an 8-bit counter and an 8-bit timer to implement its function. The timer will interrupt every 256 input clocks and the ADC will count out 64 of those timer interrupts and then use 1 cycle (of 256 input clocks) to reset the integrator. The calculation of the sample frequency is then given as follows:

```
Sample Frequency = Input Clock Frequency / (65 * 256)
```

Since the interrupt time varies depending on how many times the integrator is high, you will spend more CPU cycles on a high input signal than a low input signal. Make sure that your project can handle the worst case. Also make sure that the clock phase is set correctly on the ADC. If your input signal is coming from another switched capacitor block, then there is a chance that you need to choose the swapped option here.

This ADC is very effective at rejecting lower frequency signals. The data sheet suggests that you set the sample frequency at 100 milliseconds to achieve the optimum rejection of 50 Hz to 60 Hz. Time needs to be allotted for the ADC to complete its full cycle before reading the data. This is done by repeated polls to the ADC to see if it has finished its conversion.

There isn't a need to flush the ADC with additional samples when changing the analog input of the ADC since the integrator is reset every sample.

ADCINC14 (only available on newer PSoC parts)

The 14-bit incremental ADC is very similar to the 12-bit incremental ADC mentioned previously. The difference is that three of the digital blocks are configured to form a 24-bit PWM generator that outputs a high signal during the time of integration and a low signal during the time of calculation.

This data sheet gives an explanation for choosing the sample rate to reject unwanted frequencies. That can be applied to the other incremental ADCs. It tells us to select a sample time that is an integral multiple of the undesired frequencies. The data sheet gives the example of choosing 100 msec to reject 50 Hz and 60 Hz noise levels. A multiplication of $5 \times 1/50$ Hz and $6 \times 1/60$ Hz will give us the 100 msec interval, so both 50 Hz and 60 Hz are integral multiples that equal the interval.

ADCINCVR

The 7- to 13-bit variable resolution incremental ADC allows you to adjust the resolution of the ADC. However, if you plan to just use it in place of the 12-bit incremental ADC then you will be paying a premium of 100 bytes of Flash for the privilege. Since you can turn the resolution down to 7 bits, you are able to return up to 10,000 samples per second and still retain the advantages of the incremental ADC. It will still cost you three digital blocks though. As with the other incremental ADCs, you will need to expect some CPU overhead to complete its conversions.

DELSIG8

The delta-sigma 8-bit ADC offers a much quicker conversion rate, returning samples at up to speeds of 32,000 samples per second. Finished conversions can be set to trigger an interrupt, or you can poll the ADC to see if a conversion has finished.

Unlike the incremental ADCs, you must flush the delta-sigma ADC when changing from one input signal to another. The data sheet requirement is to wait for the third sample before the result can be relied on.

Although there aren't any limitations on the placement of the analog block used in this ADC, there are limitations on the placement of the digital block, as the digital block must be able to connect to the decimator resource of the PSoC.

Another important note is that there can only be one delta-sigma ADC for each configuration, as there is only one hardware decimator. The data sheet warns that multiple instances of delta-sigma ADCs may appear to work, but only one instance will be correct.

The newer PsoCs such as the 27xxx offer a second order modulator ADC, which will improve the signal to noise ratio. It does require one more switched cap block to implement the second-order modulator ADC.

DELSIG11

The delta-sigma 11-bit ADC is very similar to its 8-bit counterpart, including the option to have a second order modulator on the newer PSoCs. The extra bits of resolution for the 11-bit delta-sigma ADC require more processing time, so you are not able to get as high a sampling speed as the 8-bit delta-sigma ADC. This ADC also requires some additional Flash and RAM in order to operate. As with the 8-bit delta-sigma, you must flush the ADC when changing input signals and wait for the third conversion before getting reliable data.

DUALADC

The dual-input incremental ADC mimics the 7- to 13-bit variable resolution incremental ADC listed earlier. The dual-input feature means that they have taken two instances of the variable resolution incremental ADC and have them share the same 16-bit PWM. This results in an ADC that can sample two input signals at the same time. You share overhead CPU costs and reduce the overall requirement for digital blocks.

SAR6

The 6-bit successive approximation register is the fastest ADC in the PSoC. Using a comparator, it determines the amplitude of the signal by dividing the analog range in half until it determines the correct result. You can compare the successive approximation register to trying to guess a number from 1 to 100, picked by a friend with as few questions as possible. You would start at 50 and ask if it's higher or lower. If lower, then you would choose 25 and ask the same question again. Repeating this process will give you the answer very quickly.

The ADC needs a start call to set the power level and then you simply call the SAR6_cGetSample. Voila, the SAR6_cGetSample routine returns the result in the accumulator. You should be aware, however, that during the execution of the SAR6_cGetSample routine, the processor is stalled for several cycles of the ADC clock. This means that the processor can not perform any other functions, including interrupts, for this interval. Therefore, choose as fast a clock as possible to clock this ADC. I made the mistake once of putting this ADC to work with a PSoC low-pass filter. Since I was trying to filter a slow signal, the analog column clock was rather slow. Needless to say, my project didn't run very well. As a matter of fact it didn't run at all since my general-purpose timing interrupt interval was shorter than the time to complete one conversion cycle.

TRIADC

The TRIADC is the three-input version of the ADCINCVR. It allows you to sample three inputs simultaneously with the select resolution of 7 to 13 bits. The ADC shares the 16-bit PWM for all three inputs. Therefore the additional inputs only require two more digital and two more switched cap blocks in order to be implemented.

Amplifiers

AMPINV

The AMPINV is an inverting amplifier that is very similar to the PGA or programmable gain amplifier. One significant difference that you should be aware of during the planning stage of your project is that the input to the inverting amplifier must come from a switched cap or continuous time analog block. You can not connect a port pin directly to the inverting amplifier input.

Gain values will have a slightly smaller range than the PGA since the inverting amplifier gain is based on –Rb/Ra and the noninverting amplifier has a gain based on 1 + Rb/Ra. The inverting amplifier is also always centered around AGND. If you want to center it around a particular level, then that must be taken care of in another analog block or you will need to supply that level as an external AGND reference.

CMPPRG

The CMPPRG is a programmable threshold comparator that operates in a continuous time analog block. The threshold is a factor of the chosen RefValue and the Low Limit. The equation is given below. The newer parts add two more levels due to the extended gain of the continuous time blocks.

$$V_{Threshold} = Low\, Limit + \left(Vdd - Low\, Limit\right) \cdot RefValue$$

The output of the comparator can be passed through the analog output buffer to an external pin. The output also can be tied to the comparator bus to be used as an interrupt source or as an input to digital blocks.

INSAMP

The INSAMP is an instrumentation amplifier that allows you to amplify the difference between two signals. This allows you to eliminate noise that is common to both signals. The classic instrumentation amplifier requires two continuous time blocks that must be adjacent to each other in order to connect. The instrumentation amplifier also only has a gain from two to sixteen, with eight possible gain settings. If more gain is required, then you will need to place a PGA in another continuous time analog block to amplify the output of the instrumentation amplifier.

Another way to increase the gain is to use a new topology for the instrumentation amplifier available only in the newer parts such as the 27xxx. This newer topology utilizes three analog blocks; two continuous time, and one switched cap block. The newer topology allows you to have gain settings from 0.0313 to 93. This gain range is divided in 18 steps. The newer topology is also better able to eliminate noise that is present on both inputs. This performance is measured as the common mode rejection ratio.

Since you are using the switched capacitor in this gain circuit, remember that the analog column clock frequency affects the operation of the switched cap block. That frequency must be high enough for desired signals to pass. The switched cap frequency will have a low-pass filter type affect on the three-block instrumentation amplifier.

PGA

The PGA, or programmable gain amplifier, amplifies the input signal in relation to the reference signal. The newer PSoC analog blocks allow the gain to reach 48 in this module, in addition to allowing a stronger attenuation of signals. The PGA resides in a continuous time analog block and is therefore not subject to the limitations of the column clock frequency, unless the signal input is chosen to be the output of one of the switched capacitor blocks. It then inherits the characteristics of the signal source.

The programmable gain amplifier may have its output tied to the analog output buffer to send the amplified signal out to an external pin. Remember that the output buffer can only output Vss–Vdd and therefore may not be able to output the correct voltage amplification of all input signals. There was also an increased current draw in the older parts as the output buffers generated voltages close to Vdd or Vss. This increased current draw can be enough to affect a small signal project with a weaker power supply.

Analog Comm

DTMF Dialer

The DTMF or dual-tone multiple frequency generator allows you to generate dual-tone signals. These dual-tone signals can be used to pass information the same as with a touch tone phone. The tones are generated using a look up table for the values to produce a sine wave through a 6-bit digital-to-analog converter. The speed at which you index through the table determines the output frequency of the sine wave. The DTMF dialer module uses a digital block to control the timing and a switched cap block to perform the DAC function.

The size of the table and the API routines make this block a large user of Flash memory. It requires over 700 bytes of Flash, which can be restrictive, especially if your project is targeting one of the smaller PSoC parts.

Counters

Counter8, Counter16, Counter24, Counter32

The 8-, 16-, 24-, and 32-bit counters require one, two, three, or four digital blocks respectively. The counters with more than one digital block daisy chain the output and input signals of the digital block to create larger counters. This requires the modules to be placed in consecutive blocks in order to operate properly. It also means that more Flash memory is required for the larger counters, as you need to manipulate more registers.

The counters in the PSoC are down counters with period and compare values. If you are using counters larger than 8 bits and your values exceed 8 bits, these values are split up as needed in the different blocks. The counter will count down until the compare condition is true. Once true, it will assert the compare output. You are also able to trigger an interrupt when the compare condition becomes true or when the counter is reset with the period value.

The formula for the OutputPeriod is given in the following formula: OutputPeriod = (PeriodValue + 1) * ClockPeriod. This indicates that the digital blocks reload on a carry condition rather than a zero condition. This serves as an advantage as it allows you to divide the clock by 1 to 256 rather than 1 to 255. It also eliminates a divide by 0 condition, which would be considered an error.

DACs

DAC6, DAC8, DAC9,

The digital-to-analog converters allow you to set a specific output voltage with switched cap blocks. These signals can also be sent to the analog output buffers to be used for signals needed outside the chip, similar to the DTMF module.

The 6-bit version can be placed in any of the switched cap blocks. However, when you use the 8-bit or 9-bit versions they require two adjacent switched cap blocks. The output must come out of the type A or type C block which will dictate which analog column you can output the DAC on when placing in blocks that aren't in the same column. This limitation on the placement helps assure the best performance possible.

You should note that the DAC operates in conjunction with the reference voltage level. Therefore, if you want the output range to be from Vdd to Vss, then you need to set the ref mux to Vdd/2 ± Vdd/2.

MDAC6, MDAC8

The multiplying digital-to-analog converter is offered in both 6-bit and 8-bit versions. The 8-bit version requires two switched cap blocks where the 6-bit only requires one. As with the DAC module, the output of the 8-bit MDAC will be available on the type A or type C block. This block operates exactly like the DAC blocks, but you are given the option to choose the reference voltage that will be multiplied at the output stage. The DAC blocks use the REFHI signal.

Since this DAC is multiplying the input voltage, you can say that it is similar to the PGA block. However, the MDAC is implemented in switched cap blocks. You can use this to add an additional boost to the output of a filter or use the direct inputs to the switched cap blocks to amplify or attenuate a signal on external pins that can be connected directly into switched cap blocks.

Since the multiplication is done by the switched cap blocks, the output is updated at a rate proportional to the clock frequency for the analog column(s). Make sure that you have a fast enough clock to meet your project needs.

Digital Comm

CRC16

The 16-bit cyclical redundancy check module allows you to implement a cyclical redundancy check (CRC) on an input stream of data. The module allows you to set the seed value and the polynomial used. The input stream is fed into the input of the blocks with the option to invert the input. Once the CRC has been calculated, you read the result back by reading the seed register. The data sheet notes that you shouldn't stop the block while reading the value as you may get undesired results.

I²CHW (Only available on newer parts)

The I²C hardware implementation is only available on the newer parts such as the 27xxx. This hardware implementation allows you to implement an I²C slave or an I²C master. By using the hardware implementation, you can achieve a transfer rate of 400 kBits/sec, since the bit by bit transfers are happening in hardware. However, you will have to consider that you may not be able to pass bytes to the I²C hardware fast enough to sustain a continuous transfer.

The I²C hardware option requires you to use the dedicated I/O pins (P1[5] and P1[7] on the CY827443) that connect to the I²C circuitry. These are shown as being used once you place the module in the Device Editor Configuration/Pinout View. For other modules that don't require digital or analog blocks, placement may not be required, e.g., the analog mux module. However, you will not be able to use the I²C hardware block unless you place the module. This is done as you would place any other module, but you will not see the targeted digital or analog block highlighted. Once placed, the options for the module will be shown.

Note that the parameter for the address is actually shifted after it is passed to the module. Therefore an address of 0x48 will actually become 0x90 for writing and 0x91 for reading.

I²CM

The I²C master block will implement the I²C master function in firmware. Since everything is toggled via firmware, the overhead to the processor is much greater and your transfer speed is thereby limited to 100 kBits/second. The firmware implementation allows you to select any port to implement the clock and data pins, but both pins must be on the same port.

I²CS

The firmware implementation of an I²C slave is not available on the newer parts, but does still show up as available on the older parts. I believe the intent is to discourage its use where the hardware implementation is available. The amount of Flash required for the module is quite high. Also, the implementation of a slave using firmware is very difficult because you have to be extremely fast to assert the data pin or read its information when the clock transitions. To facilitate this, the I²C block becomes interrupt driven. When the transfer is taking place, you will have to plan on the processor handling that transaction and doing nothing else. You should also run the processor at 24 MHz to allow the slave to operate at the maximum frequency. This requires a tight spec voltage supply to guarantee correct operation.

IrDARX

The IrDA receive module requires some external connections to your micro to bridge different blocks together. It will use three blocks on the 25xxx/26xxx parts as it needs to have an inverter implemented. The improved interconnect logic of the newer parts satisfies this need thereby eliminating one block for the implementation of the IrDARX.

The IrDA receiver will look for RS-232 type communication, where the transitions occur in the center of a bit as a pulse, rather than set logic levels. The pulses signify zeros and there is no signal sent to signify the 1's. This is conducive to communication with a blinking light as you have with IrDA hardware.

Note: This module requires that you modify a mask register when altering other I/O pins of the same port, as this module uses a pull-up setting.

IrDATX

The IrDA transmitter block only requires only two digital blocks for its implementation. It provides the necessary pulse in the middle of the 0 transmissions. The pulse is an active low, so you will want to set your infrared LED to sink into the PSoC. The width of the generated pulse is 1/8 of the bit width.

Note: This module requires that you modify a mask register when altering other I/O pins of the same port, as this module uses a pull-down setting.

RX8

The 8-bit receiver block allows you to receive NRZ serial as is used in RS-232. The module only requires one digital communication block in order to be implemented. The RX8 will receive data sent at one eighth the clock frequency fed to the block. Since the internal frequencies aren't nice divisions of the typical RS-232 frequencies that you see on your computer, you will have to deal with some potential error if you are trying to match a computer's baud rates. Quite often, you'll also have to use another digital block to generate the desired baud rate clock needed for your receiver block.

The oversampling of the input data (x8) is used solely to center on communication. The data is only checked once in the middle of the bit to find out the value that should be latched into the shift register. There is no error checking for communication that changes states during the duration of the bit, as is done on some other receiver chips.

The module provides an interrupt when data has been received. If the interrupt is not used, then you should monitor the receiver full bit to make sure that you read the information soon enough after its receipt that additional data is not shifted in on top of the old data.

SPIM

The SPI master gives you a synchronous 8-bit transmitter that allows for bidirectional transfers. The SPIM module has four different modes that select the clock polarity and on which edge of the clock data is latched. There is also an option to select whether the least significant bit or most significant bit is sent first. This assures you that you will be compatible with the plethora of SPI type devices that are available on the market today.

The SPIM can interrupt on the condition of the transmit register being empty or upon the completion of the transfer.

You should note that the SPIM module doesn't include the implementation of a slave select signal. Most SPI type devices have a slave select signal that must be asserted before communication can commence. If your design requires a slave select signal in order to operate, you must provide that signal using a standard I/O pin under your control.

SPIS

The SPI slave block also shares the four modes for clock polarity and latching edge that the master has. It includes the slave select line in its design. The slave will not respond unless this slave select signal is asserted. However, unlike some off the shelf devices, this module does not tri-state the MISO pin when the device is not selected. If you plan on having multiple devices on the same bus with this module, then you will need to add the needed code to tri-state the MISO pin as required.

In order to send bytes back to the master, you simply load the transmit register with the next byte that will be asked for by the master. The transfer will commence as controlled by the master of the SPI bus.

TX8

The 8-bit transmitter module provides you a NRZ, RS232 type serial transmitter. It also requires a clock signal that is eight times the expected bit rate. The transmitter can provide an interrupt when the transmitter is empty to facilitate quick transactions.

UART

The universal asynchronous receiver/transmitter (UART) module is a combination of the RX8 and TX8 modules. Using the UART block also gives you the option of including the high level API for serial communications. The high level API requires several hundred bytes of Flash, but it will implement a receive buffer and can be used to look for a command structure. This high level API is stored in the uart_1plus.asm file (assuming you keep the default instance name).

The command structure is set up with the idea that a specific character is used as a terminator which will follow any command sent. Multiple parameters sent within a command are separated with a specific parameter delimiter. Once a command has been received, you are able to use such high level calls as to return the first parameter, all the parameters, or just query how many parameters have been received. The implementation of the UART plus APIs are great to interface with HyperTerminal as you can set the command terminator to be a carriage return or line feed. This allows the PSoC to respond as soon you press the **Enter** key within HyperTerminal.

Filters

The implementation of filters within the PSoC with no external components and the ability to change those filters on the fly is a defining characteristic of the mixed-signal analog array that simply isn't available in most microcontrollers. The successful use of the filter may take a little planning as well as a little trial and error. However, once mastered, the filters can allow you to make your design more robust and reliable by eliminating unwanted signals before additional processing is required.

Once a filter is placed in your project, you can right-click on the module and choose the filter design wizard. The filter design wizard helps you to calculate what capacitor values to choose and what clock is needed to achieve the desired filter results. Instructions on how to use the wizard are at the top of the pop-up window. The closer that you can match your entered capacitor values to the calculated values, the closer the filter will operate to the values that you have specified at the top of the design wizard. The values that are not open to change will turn red if PSoC Designer feels that they are getting too far away from the set value. Consider modifying the parameters of your filter in this case to make sure that you are designing a filter that PSoC can feasibly accomplish. Note that the pop-up window acts similarly to Microsoft® Excel. You will need to exit an entry field in order for the calculations to update, so make sure that you have clicked outside of the field that you just modified to see the new values.

One of the important things to note about designing a filter is that the cutoff frequency of the filter is based on the clock frequency being sent to the analog columns that contain the filter. The filter design wizard does not set up resources to generate this frequency for you. It simply tells you what frequency is important. The specified frequency must be sent to all switched cap blocks used in the filter. You will be required to create a clock using digital resources to send to the analog column. This clock can come from timers, counters, and so on. If the clock is attainable by using the VC1, VC2, or VC3 (or 24V1, 24V2) signals, you can reserve your digital blocks for other uses.

Since there are limits on what clocking frequencies can be set to the blocks, there are limits on what frequencies you are able to choose for your filters. That is the reason why you won't find a high-pass filter design in the PSoC. Unfortunately, this is the nature of switched capacitor designs. If filters that exceed these capabilities are needed in your designs, then the filter must be handled by external circuitry.

BPF2

The band-pass filter module allows you to build a module that will filter out unwanted low frequencies and unwanted high frequencies. This module requires two switched cap blocks. When you double-click on the BPF2 module from the Device Editor, user module selection view to add the filter to your project, it will pop up a window to ask you to select a topology. There are three different topologies available for the 25xxx/26xxx parts and four topologies available for the newer parts that have the enhanced analog.

The BPF2V is a vertical topology that is available on both parts, but limits placement to be in only two columns. It has one placement that will allow a direct input coming in from Port2[0]. The BPF2VA topology, available in the parts with the enhanced analog, allows you to place a vertical filter in any of the four analog columns. The BPF2VA also allows you to create better filters by combining two filters together using internal connections, plus it adds placements to make use of the direct connections from either Port2[2] or Port2[1] directly into switched cap blocks. The vertical topology requires you to send only the needed analog clock down one analog column. This gives you the freedom of only affecting one analog clock and leaving the other column clocks at levels that may work better for your ADC, DAC, or other modules.

The BPF2A and BPF2B horizontal topologies are available on both newer PSoC devices and older PSoC devices, and allow the newer parts some more flexibility in interconnections. The BPF2A is set up such that you can form higher order filters using internal connections. The BPF2B requires some off chip routing to accomplish this. In the older PSoC devices, the BPF2B is the only horizontal topology to allow direct access to the Port2[3] and Port2[0] pins. The newer PSoC devices allow a direct access to Port2[3] and Port2[1] with the BPF2A topology. Note that the horizontal topologies require the analog clock to be sent down two columns rather than one. In addition, all filters in the same column must now use the same clock.

LPF2

The low-pass filter module allows you to filter out unwanted low end frequencies. The various topologies available for the low-pass filter modules are identical to the topologies available on the band-pass filter modules.

Generic

SCBLOCK

The generic switched cap module allows you to modify all parameters of a switched cap block to create your own custom function. Note that there are differences between the different types of switched cap blocks that can work to your benefit or detriment. Since all parameters are under your manual control, the API is set up only to control power going to the block. This block is intended for the ambitious that have an understanding of op-amp design and switched cap technology.

Miscellaneous Digital

DigInv

The digital inverter provides a way to invert any signal within the PSoC. I used the inverter on older projects to get the input or output of UART communication in the correct orientation. However, the newer parts allow some inverting in the interconnect logic or just previous to entering the block itself. This ability helps to eliminate the use of an entire block just to invert a signal.

If you are inverting a signal that is brought in from a global bus, you will have a speed limitation of 12 MHz. You can speed up this response if you are using the input that is tied to the previous block.

An interrupt is available on the falling edge of the input signal.

E2PROM

EEPROM Module

The Flash in the Cypress PSoC family has an endurance of 50,000 erase write cycles. The block size of the Flash is 64 bytes. This makes the PSoC Flash a viable EEPROM solution. The Flash must be erased in whole blocks, but can be written back to 1 byte at a time. The endurance is also smaller than the expected 1,000,000 cycles of the EEPROM memory that you would find in devices like a 24C02.

Memory devices like the Flash in the PSoC have a logic state of all ones or all zeros when erased. Programming the memory devices will apply a voltage to the memory cell and change its logical state from a one to a zero or from a zero to a one depending on its erased state. You can use this knowledge of the Flash architecture to your advantage to make the most of your Flash usage.

The EEPROM module doesn't require the use of any digital or analog blocks. It is important, however, to realize the needed space in ROM(Flash) and RAM. The library to write to Flash requires 439 bytes just for the routines. Then there is a 16 byte requirement for each instance of the EEPROM module in addition to the size of the memory block that you are going to save. The RAM usage seems to have varied some with different PSoC versions. One data sheet was reporting the usage of RAM at up to 104 bytes in addition to the 8 additional bytes at the top of RAM.

The Flash is comprised of blocks that are 64 bytes in length. If you want to write to Flash, you will need to erase 64 bytes at a time. If you don't desire to change all of the 64 bytes within a block, the routines will read the entire 64 bytes into RAM, change the bytes that you want and then write the entire 64 bytes back out to Flash. This is the reason that you can top 100 bytes of RAM usage in this type of routine. If you specify that you want to write 64 bytes in your routine, then there is no reason to keep the copy of the previous contents of the Flash block since you will be overwriting every byte within the block anyway. Therefore, unless you are really pressed on ROM space in your project, I would suggest that you only use the 64-byte write to Flash to save on your RAM usage.

The 8 bytes at the top of RAM that are mentioned are at addresses 0xF8 to 0xFF. These bytes are used in the process of the routine. It's important that you don't overwrite these while the EEPROM routine is in process, and you should check to make sure that your stack will not reach that area during the EEPROM write.

Because of the complexity of setting up the EEPROM module and its associated routine calls, you may be tempted to create your own routines to write to Flash. Cypress doesn't recommend this and has been helpful on working on smaller routines to write to Flash that have some expected limitations in them. If you need to read from Flash, I recommend that you use the ROMX instruction. It's very straightforward and simple and doesn't require the complex setup that the EEPROM module needs for reading.

LCD

The LCD module is another collection of routines that doesn't require any digital or analog blocks to implement. It is designed to talk to the Hitachi HD44780 LCD module controller. This is a standard communication protocol that works with a variety of modules and may very well work with the one you get from your local electronics store. The LCD module requires 622 bytes of ROM to implement. This includes some routines to run a bar graph. If you don't want to use the bar graph capabilities, then you can save 200 bytes of ROM usage in the module by disabling the bar graph capabilities.

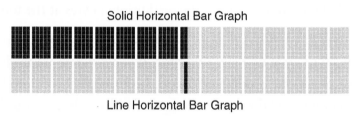

Figure 6-1: Examples of LCD Bar Graphs

The LCD module uses the 4-bit addressing method to talk to the module in order to save on I/O pins, thereby limiting the I/O requirement to seven pins. The final pin of the port is left open for your use if you desire, however you must modify the mask register created by the module when you alter the port data register. This is to allow the pull-up and pull-down options to work correctly on the port when used.

MUXs

AMUX4

The four-input analog multiplexer does not use any digital or analog blocks. It simply builds some modularized routines to control the input multiplexers of the analog columns. There is an AMUX4_Start routine although it doesn't do anything. It would be a good practice to call the routine in case future versions of PSoC Designer decide to add some code there. If you call the routine now, your project will take advantage of the improvements with the next spin of PSoC Designer without you having to remember to check and add the call in.

Ref Mux

The reference multiplexer allows you to output one of three reference signals from a continuous time analog block. The reference signals provided are sent to the output of the analog block via the test mux. The test mux is an analog path that will route the input voltage to the continuous time analog block to the output of the block without going through the operational amplifier within the block. This allows you to send an input signal to the output without having the electrical characteristics that may be imposed by the op-amp. For example, with some projects that used the 25xxx/26xxx series parts, Cypress suggested that I use the test mux to allow me to send a rail-to-rail signal from the input pin directly to the ADC.

PWMs

PWM8, PWM16

The PWM modules use one digital block for an 8-bit PWM or two blocks for a 16-bit PWM. The APIs allow you to set a period and duty cycle for the blocks. When using a digital timing block for its duty cycle capabilities, you should note that the different compare conditions change the formula for the duty cycle. The formula for the duty cycle is given in the data sheet.

$$DutyCycle = \begin{cases} \dfrac{CompareValue}{PeriodValue+1}, \text{ For } \textit{Less Than} \text{ comparison} \\ \dfrac{CompareValue+1}{PeriodValue+1}, \text{ For } \textit{Less Than Or Equal To} \text{ comparison} \end{cases} \qquad \text{Formula 6-1}$$

For example, if you set your comparison to be a less than or equal to comparison, then you will be outputting a small duty cycle even if your compare register is 0.

The data sheet also gives some special conditions that should be noted. These are conditions where it may not be apparent what state the output should be in. You should make sure that your project can handle the special conditions noted below or you will need to add in code protections to prevent the counter from being loaded with these conditions.

Table 6-1

Period Register Value	Compare Type	Compare Register Value	Ratio of Pulse-Width High Timer to Period
0	Don't Care	>0	1.0
0	<=	0	1.0
0	<	0	0.0
>0	<=	0	1/(Period + 1)
>0	<	0	0.0
Period=Compare	<=	Period = Compare	1.0
Period=Compare	<	Period = Compare	Period/(Period + 1)
Compare Value> Period	Don't Care	Compare Value > Period	1.0

If you want to stop the PWM and force the output low, then you will need to call the stop routine with the API. If you set up interconnects simply to disable the enable input, then the count stops and the PWM is left in the last state that it was while the enable bit was still active, which could be a high state.

Care should also be taken when changing the duty cycle, as you can potentially cause an unexpected glitch. If this glitch is important in your project, then you may want to reload the compare register just after a terminal count event.

PWMDB8, PWMDB16

The PWM dead band modules require two digital blocks for an 8-bit module or three digital blocks for a 16-bit module. The PWM portion of the PWMDB module is always set up to be less than or equal to, so you will need to adjust your PWM accordingly. These modules have programmable dead band time and a kill input that will drive both of the outputs low when active. The PWMDB modules are available on both the older 25xxx/26xxx parts as well as the newer PSoC families. The older parts have a larger Flash requirement to implement the blocks.

A dead band PWM is used for two quadrant controllers where you control two switching devices. The two devices are meant to turn on opposite of each other. However, you don't ever want both devices to be on at the same time. So you will have a small time where both devices are inactive while switching from one device to the other device. This is known as the "dead time." The dead time is the value of the register +1 so you will always have some dead time when using this module. The normal PWM output is available. It can be used for syncing or used in conjunction with one of the phase outputs to get the deadband signal on only one of the two edges.

The kill input is meant to drive both outputs low when active and is available on PWMDB modules that are created in both the older families of PSoC and in the newer families. However, the newer PSoC parts add the kill mode parameter. An explanation is given in the data sheet for the PSoC part.

Synchronous Restart Kill Mode – Internal state is reset and reference edges are ignored until the kill signal is negated.

Disable Kill Mode – Kill signal must be negated and user must re-enable the block in firmware to resume operation.

Asychronous Kill Mode – Outputs are low only for the duration that the kill signal is asserted, subject to a minimum disable time between one-half to one and one-half clock cycles. Internal state is unaffected.

You should review the timing waves within the data sheet for a more detailed understanding of the kill input's effect and the timing of the dead time.

Random Sequence

PRS8, PRS16, PRS24, PRS32

The pseudo-random sequence (PRS) modules will allow you to produce a pseudo random bit stream. *Pseudo* is a Greek-based prefix that indicates that the random sequence isn't truly random, but it appears to be. The number of bits involved in the random sequence generator will change how long of sequence will be generated before the sequence begins to repeat.

The operation of the PRS will shift numbers through a shift register. It will use bits from specific locations within the register in an XOR equation to generate the next bit shifted into the register. You will need to provide a seed value for the register and a polynomial which specifies which bits you will use in the XOR equation.

Different polynomials will have different effects on the output stream. The PRS data sheet has a table showing a suggested polynomial for various bit lengths. The table isn't intended to show a complete list of all polynomials that would have acceptable characteristics, but simply to show an example polynomial that would have generally desired characteristics.

Temperature

FlashTemp

The Flash temperature module uses one analog switched cap block to read the temperature of the PSoC. The temperature is returned by the API where 1 count = 1 degree Celsius. Note that this does not necessarily match the temperature of other areas of your PCB or the temperature of the surrounding air.

The Flash temperature module requires that you set the reference mux setting in the global parameters to be Vdd/2 ± BandGap or (2 * BandGap) ± BandGap. It also requires that you set the op-amp Bias setting to LOW. This can affect other areas of your project such as ADC blocks, so if needed, reconfigure to take the temperature reading, then reconfigure the global parameter back before starting your other analog processing.

Timers

Timer8, Timer16, Timer24, Timer32

The digital blocks of the PSoC can be used independently as an 8-bit timer or daisy-chained together to form 16-, 24-, or 32-bit timers. Larger timers require more Flash for their routines since you need to initialize and handle more registers.

The period for the timer is equal to the period register plus one times the period of the clock signal. This allows you to reach a divisor capability from 1 to 256 for each 8-bit timer. The timers allow interrupts on the terminal count or on a compare event. This works well if you want to interrupt on a specific interval as shown in the *Project Walkthrough* of Appendix B. The timer can also be used to time an external event. You need to make sure that the external signals follow the limitation of 12 MHz for external signals. The terminal count and compare functions can also be output to pins to trigger other external events within the PSoC or can be routed to external pins.

CHAPTER 7

Interconnects

Understanding the interconnects between the PSoC blocks can be your best friend in understanding how to use the PSoC microcontrollers effectively. This chapter will explore some of the basics about the interconnects and discuss some general rules for their usage. A graphical depiction of the interconnect system is visible in the interconnect view of the Device Editor. PSoC Designer allows you to hold down on the **Ctrl** key while you click the mouse to zoom in on the area the mouse is over. You can alternately zoom out by holding the **Ctrl** and **Shift** buttons together while clicking. Holding the **Alt** key will change your mouse pointer into a hand that will *grab* the view of the interconnects when you press the mouse button. You can then drag the view around the screen by moving the mouse as you continue to hold down on the mouse button. Unless you have a really large screen, this helps you zoom in and manipulate various areas of the interconnects.

All the PSoC families that were developed after the 25xxx and 26xxx parts have an interconnection system that differs significantly. Therefore, I will first talk about the interconnection system of the 25xxx and 26xxx parts and then cover the newer system in the latter half of this chapter.

25xxx/26xxx Interconnection System

Digital Interconnects

The basic layout of the digital interconnects for the 25xxx/26xxx series is depicted in Figure 7-1.

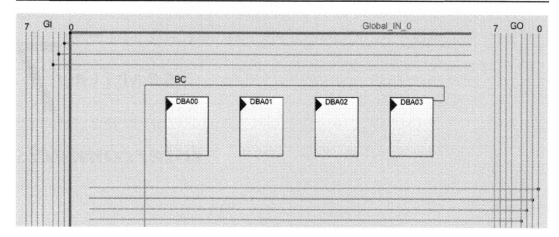

Figure 7-1: Top Row of Digital Blocks

There are two general inputs to the digital blocks: the logic input and the clock input. The clock signal is designated with the black triangle. The clock input can connect to any of the global clocking signals, the previous digital block, or various interconnects. The previous block signal source allows a chaining of multiple digital blocks for timers, counters, etc. without using the other interconnect resources. The logic input has the various interconnect settings. It doesn't allow connections to the global clocks, but does allow a logic low or a logic high connection. The logic input also allows connections to the comparator bus signals from the analog columns below.

The global buses are represented in two columns around the digital blocks. There is a label at the top of the columns depicting the organization of the buses. At the top left you will see a column labeled GI. This represents the global input lines, which are ordered from the global input bus 7 in consecutive order down to global input bus 0. At the right of the digital blocks there is another column labeled GO. This column represents the global output bus. It is also ordered from global output bus 7 down to global output bus 0. If you move your mouse over these buses, you will see that each bus will highlight in a red color and the name of the bus or float next to your mouse pointer. The screen capture of Figure 7-1 shows that I've highlighted the global input bus 0. The mouse location in the screen capture is omitted, but from the text position, you can tell that I was above the DBA03 block.

Notice that the top row of digital blocks is adjacent to row interconnects that lead to the lower nibble of the global input and global output buses. The bottom row of digital blocks is adjacent to row interconnects that lead to the higher nibble of the global input and global output lines. This illustrates the correlation between module placement and the ability to attach to the global buses. If you place a module so that its input or output resides in the top row of digital blocks, then you will only be able to connect that input or output to a lower nibble of the global bus.

You will recall from Chapter 2 – *Structure of the PSoC* that the PSoC is structured so that the global bus line's bit designation dictates the port bits that it can connect to. Following is an example. I've placed an 8-bit timer module in digital block DBA03 as shown in Figure 7-2. From the user module parameters area, I've selected the clock input to come from Global_IN_1. This is bit 1 of the global input bus. I've selected the output to connect to Global_Output_0. When you make that selection, you will see a line drawn in the interconnect view that shows the attachment from the timer module to the respective global lines.

Once the user parameters were selected, I set up the port pins by selecting Global_IN_1 for Port_0_1. This sets up the registers so that the drive mode for that pin is high impedance and the pin signal is routed through a buffer to connect to the Global_IN_1 line that is connected to my timer module. I used the same process to

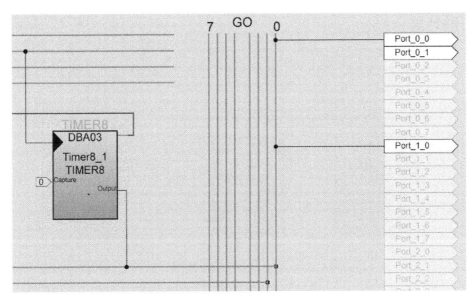

Figure 7-2: Placement of a Timer

connect both Port_0_0 and Port_1_0 to the Global_OUT_0 bus. Once the pin has been assigned as the global resource, you will notice that it changes coloring slightly in the interconnect view to indicate that this resource is currently in use.

My selection above connects two pins to the same Global_OUT_0 signal. PSoC Designer will allow this since the two signals won't be in conflict with each other. There selection to the bus is handled in the drive mode and global select registers for each port. You are not allowed to connect two signal sources to one digital bus. These signals are buffered before the global bus and could possibly conflict with each other. Please note that PSoC Designer stops you from doing this, but the possibility of making this mistake is still possible by manipulating the ports registers directly. This could cause unexpected results and damage to your micro.

The same rules apply with output signals from your digital blocks. PSoC Designer will prevent you from placing another module in a digital block and allow that module to drive the Global_OUT_0 bus. Since this bus is already being driven by Timer8_1, any other connections that would drive this bus line would be in conflict. However, you can use Global_OUT_0 as an input to other blocks as this does not cause a conflict on the line. An example of this is shown in Figure 7-3.

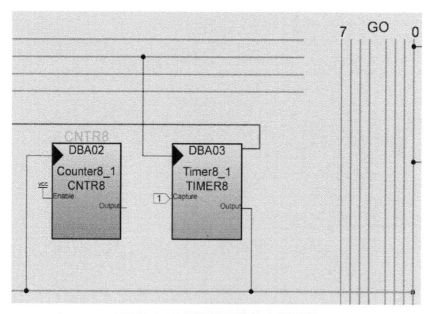

Figure 7-3: Using a Global Output Signal as an Input

The comparator bus is an interconnect that is available only to the logic input of the digital block. Certain analog modules have the ability to connect to the comparator bus and thereby provide a digital signal to the digital blocks. I specified ComparatorBus_0 for the input of the capture signal for the Timer8_1 module. A tag is placed in the interconnect view with a 0 in it showing this connection. The graphical depiction of the comparator bus is at the bottom of the interconnect view next to the analog blocks. The Cypress Semiconductor Corporation software team must have felt it a little too messy to clutter up the screen with a connection that would cross over other connection to show this selection. If I were to select comparator bus 1, 2, or 3, then the label inside the tag would indicate 1, 2, or 3 accordingly. You will notice that the high selection for the Counter8_1 input is depicted with a Vcc signal. Any signal that connects to a resource that is not in close proximity or is not visually depicted within the interconnect view will get a tag symbol when selected.

Figure 7-1 shows the DBA03 broadcast connection. It is the connection labeled BC and originates from the DBA03 digital block. This connection is available to all other digital blocks for both the clock and logic input signal. It the only connection that will allow you send a logic input signal from one digital block row to the other digital block row.

The PSoC has one other method of sending a clock signal from one row to the other internally without using the DBA03 broadcast signal. This is by using the Global_OUT_0 and Global_OUT_4 buses. The top row of digital blocks can use GLOBAL_OUT_4 as a clock signal input. The bottom row of digital blocks can use GLOBAL_OUT_0 as a clock signal input. Even though this does cost you one global signal to use this method, it doesn't require any off-chip routing.

Analog Interconnects

The analog block system layout is depicted in Figure 7-4.

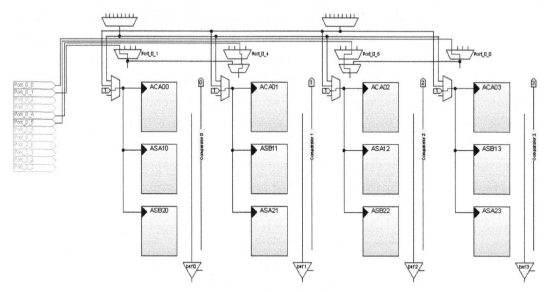

Figure 7-4: Analog Block System

The analog blocks, like the digital blocks, have input signals and clocking signals. The clocking signals are routed to the multiplexers on the top left of each column of analog blocks. These multiplexers are used to select the clock signal for the entire column. The multiplexer can choose the clock from one of four sources. The options are 24V1, 24V2, AnalogClock_0_Select, or AnalogClock_1_Select. The analog clock selects have the option of connecting to the output of any one of the digital blocks above. This allows you to use the digital blocks as timers or counters to create the exact desired frequency for the analog blocks. When the digital block is the source for the analog block's clock signal, it will be necessary to make sure the module that is providing the clock signal to the analog blocks is started and running for them to work as expected.

Directly above the analog blocks are multiplexers that allow pin inputs to be run to the analog blocks. The multiplexers are only able to connect directly to the continuous analog blocks, but those continuous analog blocks can become inputs for the switched cap analog blocks below, so I will refer to the multiplexers as inputs for the entire column of analog rather than just the continuous analog blocks. You need to select AnalogInput as the drive option for any pin that you intend to use as an analog input. These multiplexers are labeled left to right as AnalogColumn_InputMUX_0, AnalogColumn_InputMUX_1, AnalogColumn_InputMUX_2, and AnalogColumn_InputMUX_3. You should take note that the AnalogColumn_InputMUX_1 and AnalogColumn_InputMUX_2 each have an additional multiplexer directly beneath them. They are referred to as AnalogColumn_InputSelect_1 and AnalogColumn_InputSelect_2.

Each input multiplexer only allows connections to four different inputs. The only port that these multiplexers connect to is port 0. The interconnect view shows these port pins to the left of the analog blocks which allows a visual depiction of when you make this connection. Select each multiplexer by click on it to see what connection options are available. AnalogColumn_InputMUX_0 allows connections to the odd bits of port 0, i.e., Port_0_1, Port_0_3, Port_0_5, and Port_0_7. AnalogColumn_InputMux_1 allows connections to the even bits of port 0, i.e., Port_0_0, Port_0_2, Port_0_4, and Port_0_6. The input select multiplexers allow columns 1 and 2 to select one of two multiplexers to use as their input. I've listed the possible connections in Table 7-1.

Table 7-1: Input Selection Possibilities for Analog Columns

Analog Column 0	Analog Column 1	Analog Column 2	Analog Column 3
Port_0_1	Port_0_0	Port_0_1	Port_0_0
Port_0_3	Port_0_2	Port_0_3	Port_0_2
Port_0_5	Port_0_4	Port_0_5	Port_0_4
Port_0_7	Port_0_6	Port_0_7	Port_0_6
	Port_0_1	Port_0_0	
	Port_0_3	Port_0_2	
	Port_0_5	Port_0_4	
	Port_0_7	Port_0_6	

Analog Columns 1 and 2 have the potential to connect to any bit of port 0, however, if they choose to select the multiplexer that isn't directly above their column, then they must share the input selection of the other column. For example, if Analog Column 0 needs to be connected to Port_0_1, then Analog Column 1 cannot be using Port_0_3 at the same time. It would have to use Port_0_1 if you select AnalogColumn_InputMux_0 as its input.

The analog input multiplexers are some of the first connections that I found the need to change during operation. They were used to select different channels of input signals for my analog-to-digital converters. These multiplexers are controlled at runtime by the analog input select register (AMX_IN) and the analog output buffer control register (ABF_CR). If you use the AMUX4 user module, you can select between the different analog input select register options using its API calls. There won't be a need for you to manipulate the register directly. However, this only chooses the values for the column input multiplexers. There is no API to allow you to switch the input switches that determine which multiplexers column 1 and column 2 are going to use. That option is stored, as noted earlier, in the ABF_CR register. This register is a write-only register on older parts, which means that you will not be able to perform AND or OR instructions to manipulate one bit only. You will need to keep track of the proper state of the other bits within the register and use a MOV command to load the register. I recommend that you examine the PSoCConfigTBL.asm file to see what state PSoC Designer is loading into all the bits of that register and then adjust your MOV command accordingly. The ABF_CR register resides in bank 1 of RAM, so you will be required to switch banks to bank 1 in order to load the register properly.

The comparator bus for each row of analog lies to the right of each column. It terminates in a yellow numbered label that matches the style of label shown in Figure 7-3. The comparator bus is a logic signal that can be used with the digital blocks. The actual state of the comparator bus can be fed through a digital block to an outside pin, or simply used to control the digital blocks. Only one analog block can feed any comparator bus at any one time.

The analog output bus also lies to the right of each column. The analog output bus can be used to send an analog output signal to an external pin via the analog output buffer at the bottom end of the bus. The analog output buffers start to draw more current as they approach the rails, so you may want to use the comparator bus to pass the state of a switching signal. The analog bus can also be used as a method to send the signal output from a continuous time block to a lower switched cap block as the switched cap blocks have some input options which use the analog output bus. You can also use the bus signal as an input for the continuous analog blocks. One particular example that I used was to enable the test mux of the continuous analog block to pass the port pin signal directly to the analog output bus without going through the op-amp. This allowed me to pass a rail-to-rail signal to the analog output bus where it was used by the ADC to convert a rail-to-rail signal.

The lower nibble of Port 2 can be used to input analog signals to particular switched cap blocks. Port 2 bit 0 can be sent to the BMux of ASA23. Bit 1 can be sent to the AMux of ASB20. Bit 2 can be sent to the AMux of ASB13. Bit 3 can be sent to the BMux of ASA10.

22xxx/24xxx/27xxx/29xxx Interconnection System

When Cypress Semiconductor Corporation worked on their second generation of parts, they looked to make the most important improvements that would add value to their parts. The improvement of interconnecting signals was quite significant. Routing signals became much more powerful. They also have added some basic logic to the interconnects themselves.

Digital Interconnects

The layout of the digital interconnect system of the 27xxx is shown in Figure 7-5. It sets the basis for the principles used in the newer PSoC blocks.

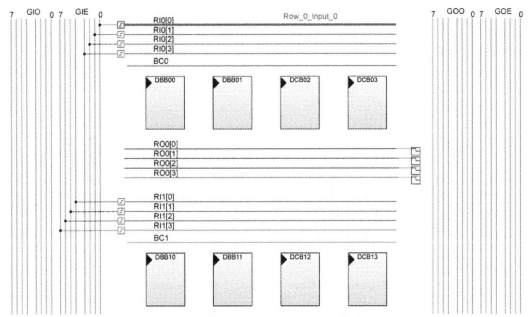

Figure 7-5: Digital Interconnects of the 27xxx

At first glance you might be thinking that this is the same system as with the 25xxx/ 26xxx parts. However, you don't have to compare the two long before you start to see some immediate differences. First of all you notice that there aren't eight global bus lines on either side, there are now 16. The labels are no longer GI (Global In) and GO (Global Out). They are now labeled GIO (Global In Odd), GIE (Global In Even), GOO (Global Out Odd), and GOE (Global Out Even). The odd and even designations are tied into the port number. Port 0 and port 2 are even ports. Port 1 and port 3 are odd ports and so on.

I've highlighted the top input row the same as I did in Figure 7-1. However, you can see that my highlight doesn't label that row and the corresponding global bus connection as the same any more. The top row now has a unique name separate from the global bus connection to the left. It is now a separate resource. The name is shown just above the line as being Row_0_Input_0. Likewise, each row is a unique resource. The light gray color is used to denote the input rows, and dark gray indicates that the row is an output row. At the end of each row is a small box symbol. These symbols represent a multiplexer or demultiplexer that allows the new variety of connections. You may note that Figure 7-5 shows some connections between the interconnects that are above the digital blocks to the left input global buses. These are the default settings for the digital interconnect and can be reassigned to different global buses. All the possible settings for a row input interconnect can be seen by clicking on that interconnect. Figure 7-6 shows an example.

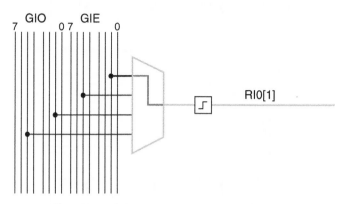

Figure 7-6: Row_0_Input_1 Multiplexer

Each input multiplexer has the ability to connect to one of four different global bus signals. There is no option to disconnect the row input from one of these four connections. You will notice that the four inputs all hold the same position within the global bus nibbles shown. Since Figure 7-6 shows row 1, it connects to position 1 in each nibble of the global input bus connections, i.e., bits 1 and 5 of both the odd and even global buses. This relationship is important to note in order to arrange module placement and pin assignments.

Since the even and odd ports are assigned to different global buses, this will allow you to use the same bit of two different ports as global inputs. For example, you could have a signal that resides on port 0 bit 0 that could be routed to the top row of digital blocks using Row_0_Input_0. You could also route a different signal on port 1 bit 1 to Row_1_Input_0. This is not possible in the 25xxx/26xxx parts. In addition, you are able to route the same I/O pin to a row signal on each row of logic by using the input multiplexers. This also requires additional resource usage or external routing to accomplish on the 25xxx/26xxx parts.

The output logic for the digital signals adds a new dimension in capability. Figure 7-7 shows a graphical description of this capability. It is shown by clicking on the

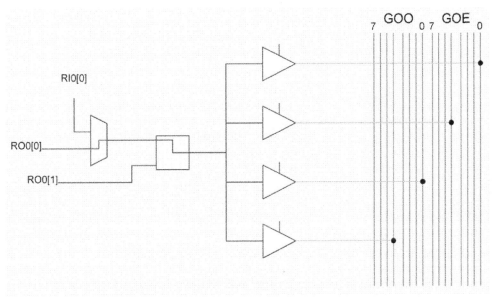

Figure 7-7: Row_0_Output_0 Demultiplexer

square at the end of the output row. First, the outputs are also improved in much the same way as the input signals. They are able to connect to one of four global buses. Like the input signals, these global bus assignments are dependent on the row value. Unlike the input signals, it is possible to have all of these four disconnected, leaving all global bus lines open for other uses. These connections are represented by the four separate latches connected to the global buses. At the far left is a two-input multiplexer that allows you to direct the signal from that row to the output or select the input signal from the corresponding input row above to the output. This allows you to set up a connection directly from one port pin to another port pin.

The square box in the middle of Figure 7-7 is a logic table that can have one of 16 different values. These values reference inputs A and B. The A input is the output of the multiplexer to the left. The B input is the other output row listed below the multiplexer. The possible values for the output of the logic table are as follows.

Table 7-2: Logic Table Values of Row Interconnects

Logic	Description
A	A only
B	B only
~A	Complement of A
~B	Complement of B
A_AND_B	Logic AND of A and B
A_NAND_B	Logic NAND of A and B
A_AND_~B	Logic AND of A and the complement of B
~A_AND_B	Logic AND of B and the complement of A
A_OR_B	Logic OR of A and B
A_NOR_B	Logic NOR of A and B
A_OR_~B	Logic OR of A and the complement of B
~A_OR_B	Logic OR of B and the complement of A
A_XOR_B	Logic XOR of A and B
A_XNOR_B	Logic XNOR of A and B
TRUE	Always a high
FALSE	Always a low

A quick check of all the other logic tables available on the digital rows show that the B input for that row comes from the output row directly underneath it. The B input for row 3 comes from row 0.

The improvements to the digital blocks also added two features that play a role in interconnects, even though their control is handled in the function register of the digital blocks. Bit 7 controls an invert capability on the data input for the digital block. The 25xxx/26xxx type parts required the use of an additional digital block in order to invert a signal. Bit 6 controls the connection of the block output to the broadcast row. Figure 7-5 shows the BC0 (Broadcast 0) and BC1 (Broadcast 1) interconnects. Each digital block is able to connect to the broadcast connection. PSoC Designer will not connect multiple blocks to the same broadcast interconnect. You should make sure that you don't inadvertently make this same mistake when modifying the function register of the digital blocks.

The broadcast interconnects are available as inputs to the other digital blocks within the same row. This allows an output signal from one block to be passed to another block in the same row without using the row output interconnects. The selection for which block feeds the interconnect signal can be made by clicking on the broadcast symbol in the interconnect view. The broadcast interconnect can also get its source signal from the other broadcast signal. You can use this option to get a signal from the output of one digital block to another digital block in a separate row without using the row output interconnects. This is similar to the DBA03 broadcast signal of the 25xxx/26xxx parts, but now you are able to use any digital block as the source signal for any other digital block.

Analog Interconnects

The analog interconnect system of the newer PSoC parts is very similar to the interconnect system of the 25xxx/26xxx parts. It has the same connection possibilities with I/O pins for the analog columns and with the input multiplexers. There is however an addition to the comparator buses of the analog blocks. Cypress has added logic tables to the top of the comparator buses. These logic tables allow you to customize the signal going to the digital blocks. This is a great addition as you are now able to set the logic in such as way as to trip the comparator interrupts when desired. These logic tables have two inputs. The B input will come from the comparator bus immediately to the right. Comparator bus 3 has its B input generated by the state of comparator bus 0.

PSoC Memory Management

PSoC Designer will try to take care of as much memory management as possible. This chapter will try to explain more about how memory is managed on the assembly level. Memory for the PSoC microcontroller is divided into two areas: RAM and ROM. RAM is an acronym for random access memory. It refers to the registers that can be accessed and changed to stored variables within your program. ROM is another acronym that stands for read only memory. The PSoC microcontroller does use Flash for its ROM which can be overwritten by the program, but the legacy acronym of ROM is used to denote this Flash program space. ROM will be used to hold program instructions, and sometimes look-up tables for the program operation.

Press **F7** while in the Application Editor of PSoC Designer, and you will initiate a **Build** of your project. PSoC Designer will use the assembler to transform your assembly or C code into the hex file that will be loaded into your chip. The assembler first converts each file individually, and then the linker will combine the various files into one file. This chapter intends to explain a few points about what is going on in this process so you can not only design cleaner projects, but you will also be alert for potential problems.

In order to understand how memory is handled, you should be familiar with the mapping file generated during a build. This mapping file will reside in the output folder of your project. The file name is the same as the name of your project and the mapping file has the extension .mp. This file will only be generated after a successful build. If you are not able to build your project, then you cannot generate a mapping file. It is important to note that it is possible for PSoC Designer to leave a previously generated mapping file in the output directory if you are simply doing a build and the compiler runs into some errors. This mapping file will not necessarily match the new mapping file that will be generated when you successfully build your project, but it has been helpful to me in the past sometimes to figure out what my problems with the current build may be. The mapping file can be of particular interest when you are trying to squeeze in the last few bytes of a program that fills all the Flash of the micro. If you have a desire always to start over fresh, you can select the **Rebuild All** option, which will start you over fresh with no mapping file until a successful code generation, or you can choose **Clean** from the **Project** menu which will delete all generated files before choosing to build your project.

Once you have built a project successfully, even if you haven't added any code of your own, you can examine the mapping file. For this chapter I will refer to the mapping file that was generated using the project bare bones that I generated during Appendix B – *Project Walkthrough*. This project uses the CY8C27443 PSoC. If you haven't already, I suggest that you enable the **Output File Tree** in the **Project Explorer** of the Application Editor. This is done by clearing the bottom checkbox of the editor options. Click on the menu **Tools**, select **Options**. Select the **Editor** tab and you will see this checkbox near the bottom of the window. After you clear the checkbox and press the OK button an **Output** tab will appear on the bottom of the Project Explorer file list at the left of the Application Editor area. If you don't see the file list then choose **Project** from the **View** menu. It allows you to get easy access to the output listing and mapping files while in Application Editor.

Areas

When you first open the mapping file you will see the word 'Area' repeated at the beginning of sections to the left of the file. The Area directive is used to specify sections of ROM and RAM. There are several areas that PSoC Designer will generate automatically. The mapping file has listed these areas. The first area of my project is included below:

```
Area                     Addr   Size   Decimal Bytes (Attributes)
---------------------    ----   ----   ------- ----- ------------
                  TOP    0000   00D0 =    208. bytes (abs,ovr,rom)

        Addr   Global Symbol
        -----  --------------------------------
        0068   __Start
        00A9   __Exit
```

The other words listed on the same line as the Area directive are column headings describing the values listed below them. On the second line, the first word that you see is the name of the area. This area, called TOP, is generated by PSoC Designer and refers to the first area of Flash memory in the PSoC. The area TOP has a beginning address of 0000. A quick reference to the PSoC data sheet shows that this area is the location of the interrupt vectors. Immediately after the address of the area, the size of that area is listed. Since the size is D0 (this is a hexadecimal number) then you know the ending address of the TOP area is CFh. They give you the size in decimal for your convenience if you are like the rest of the world and learned everything in base 10. The final listing is the attributes of the section. These attributes may be listed specifically when declaring the area or they may be the default attributes if they weren't specified. The area TOP has the attributes of an absolute address, overwriting other areas of the same name, and it resides in ROM (Flash) memory.

Immediately below these two lines are labels that are declared in the area TOP. Again we have column headings separated with dashes from the entries below. At address 0068h, there is a label called __Start. This is the beginning of coded instructions. Address 68h is immediately after the interrupt vector table at the beginning of memory. In between the __Start and __Exit labels are all the code that is executed before executing your added code in main.asm. Note that the boot section of code in the area TOP also calls other sections of code that don't reside in the TOP area. In fact the next two areas listed in my mapping file are related to such sections.

```
Area                      Addr   Size   Decimal Bytes (Attributes)
----------------------    ----   ----   ------- ----- ------------
                  lit     0100   0109 =    265. bytes (rel,con,rom)

        Addr   Global Symbol
        -----  --------------------------------
        0100   __lit_start
        0100   LoadConfigTBL_barebones_Ordered
        01AF   LoadConfigTBL_barebones_Bank0
        01E0   LoadConfigTBL_barebones_Bank1
        0209   __lit_end

Area                      Addr   Size   Decimal Bytes (Attributes)
----------------------    ----   ----   ------- ----- ------------
        psoc_config       0209   006B =    107. bytes (rel,con,rom)
```

```
Addr   Global Symbol
-----  --------------------------------
0209   _LoadConfigInit
0209   LoadConfigInit
0209   __psoc_config_start
0210   _LoadConfig_barebones
0210   LoadConfig_barebones
0274   __psoc_config_end
```

Looking at the names of the labels (routines) found within the areas lit and psoc_config, you can see that they are loading configuration tables. These routines and tables are used to initialize the control registers of the PSoC digital and analog blocks and the microcontroller core blocks in order to match the configuration selections that you have made in PSoC Designer. You will notice that these areas along with most other areas defined have the attribute 'rel,' which stands for relative. These areas do not have to reside at a particular address. The linker is able to place them at the address that makes the most sense for optimization.

You may notice that there are two entries at the same address with nearly the same label. This is due to the naming conventions decided on by Cypress for functions called in C and for functions as called in assembly. They are indeed the same routines in either case and will perform the same operations. You may also notice that this area and in some of the following areas, there is a label that ends with _start and contains the name of the area. For example, the psoc_config area just listed has a label __psoc_config_start at address 209. This address is the same as the address of the _LoadConfigInit and LoadConfigInt tables which were just established as being the same routine. The __psoc_config_start label is simply a label that denotes the start of the area. I would not recommend calling this label as a routine or use it in your code as a replacement or substitution for the _LoadConfigInit or LoadConfigInit labels, as it will be sure to confuse others and most likely yourself in a relatively short length of time.

Just below the psoc_config area, you will find the UserModules area. This area is defined to contain all the PSoC Designer generated code for the API functions of the user modules that you have selected and placed in your project. The files will not be generated for modules that have been selected, but not placed in your project. I haven't included this whole section as it gets rather lengthy very quickly.

```
Area                          Addr   Size   Decimal Bytes (Attributes)
----------------------        ----   ----   ------- ----- ------------
        UserModules           0274   0064 =    100. bytes (rel,con,rom)

    Addr   Global Symbol
    -----  --------------------------------
    0274   _Timer8_1_ISR
    0274   __usermodules_start
    0293   _Timer8_1_EnableInt
    0293   Timer8_1_EnableInt
    0297   Timer8_1_DisableInt
    0297   _Timer8_1_DisableInt
    029B   _Timer8_1_Start
    029B   Timer8_1_Start
    029F   Timer8_1_Stop
    029F   _Timer8_1_Stop
```

Looking the UserModule area, you will be able to ascertain quickly that PSoC Designer can add a lot of code rather quickly. There are many applications that can be designed without using all of the API routines that are generated by PSoC Designer. It used to be a bit of a beast to go through and remove all the unnecessary routines that weren't being used when the project required that space to be recovered, but in the latest editions of PSoC Designer (4.2 and newer), you have the option to remove API routines that aren't called.

Below the UserModule area is the text area. The text area will include the main. asm file and other files that you add to your project, unless you specify a different area within that file. The text area of the mapping file is organized just like the other areas. In the text area, you will find all the labels that you have defined in the area specified as text or that are defined in files with no area designation and that are global. A label is global if it is called out using the export directive, or if the label is followed by two colons rather than one.

```
Area                          Addr   Size   Decimal Bytes (Attributes)
----------------------        ----   ----   ------- ----- ------------
                text          02D8   001C =     28. bytes (rel,con,rom)

    Addr   Global Symbol
    -----  --------------------------------
    02D8   _main
    02D8   __text_start
```

```
02DE  mainloop
02F4  __text_end
```

Following the text area, we see the areas of RAM that are defined in your project. The first area is named virtual_registers. This area is used if there is a need to have registers that shadow particular control registers. One instance where this would be applicable is if you were to use the LCD tool box module. Here the LCD tool box module will need to have shadow registers of the drive mode and data registers. Since the LCD tool box module uses the 4-bit method for talking to the LCD driver it only requires eight I/O pins to work. These I/O pins will all come from the same port. Shadow registers are employed so that the API calls can change the pins of the selected port without affecting the state of the unused I/O pin in the case that the unused I/O pin is configured as a pull-up or pull-down. Possible ways that the pin could be affected were discussed in Chapter 2 – *Structure of the PSoC*. I have included the example virtual_registers area from a different project to give you an example of how this section may look when using a LCD tool box module.

```
Area                          Addr   Size   Decimal Bytes (Attributes)
-----------------------       ----   ----   ------- ----- ------------
        virtual_registers     0000   0004 =      4. bytes (rel,con,ram)

        Addr   Global Symbol
        -----  --------------------------------
        0000   _Port_3_Data_SHADE
        0000   NO_SHADOW
        0000   _NO_SHADOW
        0000   Port_3_Data_SHADE
        0001   Port_3_DriveMode_0_SHADE
        0001   _Port_3_DriveMode_0_SHADE
        0002   Port_3_DriveMode_1_SHADE
        0002   _Port_3_DriveMode_1_SHADE
        0003   __r0
        0004   __virtual_registers_end
```

The __r0 register is used temporarily during the execution of boot.asm. It should be free to use during your segments of code called from main.asm, but use of the blk directive will not assign your RAM to fill that same location.

The final area listed in my basic project is called *bss*. I've included the section as it appears in my bare bones project of Appendix B.

```
Area                        Addr   Size   Decimal Bytes (Attributes)
----------------------      ----   ----   ------- ----- -------------
                     bss    0001   0003 =     3. bytes (rel,con,ram)

      Addr   Global Symbol
      -----  --------------------------------
      0001   __bss_start
      0001   timeflags
      0002   tsecond_count
      0003   second_count
      0004   __bss_end
```

This area is first declared in boot.asm. The comment listed just above the declaration declares the area to be for general use.

```
;------------------------------
;   RAM segment for general use
;------------------------------
  AREA bss(ram,rel,con)
__bss_start:
```

Since the boot.asm file is regenerated every time you use the generate application file, I wouldn't recommend that you declare your general-purpose RAM within the boot.asm file. I recommend that you declare global variables in a file you create for that purpose. The bare bones project declaration is included below. It appears in a file called ram.asm, that I created. Notice that there is a double colon after the name, which is the equivalent of the using the export directive so the reference can been seen by other files in the project. Note that it's not recommended to use the same name for both a global variable and a local variable.

Local variables that you want to be private to a file can be declared in that file. You will need to declare the text area after the bss area so that your source code for the file is assembled and linked properly.

Area bss(RAM)

```
    timeflags::       blk   1     ;register divided into various flags
    tsecond_count::   blk   1     ;counter for a tenth of a second
    second_count::    blk   1     ;counter for a second
```

Where Does My RAM Go?

PSoC Designer is designed to handle RAM management to allow the user freedom from deciding where everything needs to go. The suggested method for allocating RAM for use in your project is to use the BLK directive as used in the example code earlier in this chapter. The BLK directive will allocate bytes of RAM for use in your project. The correct usage is to list a label followed by the BLK directive. The final argument is an integer specifying how many bytes of RAM to allocate.

MyVariable: BLK 4

This allocates 4 bytes of RAM for my variable. This statement is to reside in source code files and not in include files in order to allocate RAM as desired. It must also reside within an AREA statement that points to the RAM of the project.

PSoC Designer will try to place all RAM at the lowest address possible, as the stack for the Cypress PSoC will grow towards a higher address. There is an exception with the 25xxx/26xxx families, where the stack needs to be placed at address 40h or higher to guarantee correct operation if a receiver is present. This exception is not needed on the newer PSoC parts.

The expected location of the stack can be seen in the mapping file.

```
Area                        Addr   Size   Decimal Bytes (Attributes)
-----------------------     ----   ----   ------- ----- ------------
            Stack_Zone      0004   0010 =     16. bytes (rel,con,ram)

    Addr   Global Symbol
    -----  -------------------------------
    0004   _stack_start
    0014   __Stack_Zone_end
```

The _stack_start location is determined by the end of the bss area. If you decide to declare ram areas of other names they may fall after the Stack_Zone area. By default the Stack_Zone area is 16 bytes in size. If you plan on using areas with other names and their placement follows the Stack_Zone area, then you should take care to make sure that the stack will not grow out of the reserved size. There is no run-time protection that keeps the stack from growing too large and therefore growing into your other RAM areas.

If you have declared enough RAM to place the end of the Stack_Zone at an address higher than FFh, then an error will be generated to show that the stack does not have enough room. As of yet, I haven't discovered where you are able to change the size of the Stack_Zone.

Static Allocation of RAM

Some projects benefit from having the RAM use static locations. For example, I have had several projects where I would like a particular variable to be at the same location for each project. It allows a consistency that I can rely on to find that particular variable at a known memory location without having to examine the mapping file each time that I change from project to project to find variables that exist in every project. This has been useful for external connections that query variable values by address or some debugging screens that show variable values by address and that don't have enough code space or horsepower to call out all variables by name. Another benefit that I've personally used with the static addressing method is the creation of bit macros where I am able to give a bit a single name for reference in my files. For example, without using the static RAM method, I have been unable to successfully have a bit clear instruction that can work as follows:

```
bclr   myflag     ;clear the myflag bit
```

I am able to make that instruction work correctly with static addressing. There are some easy ways to make a bit macro that handles the bit with two names; one for the bit and one for the register. Therefore, the instruction becomes something similar to the following:

```
bclr   myflag,flags1 ;clear the myflag bit in the flags1 register
```

This is very similar to how bits are handled by most of the application notes that I've used. If I don't follow a convention where I rename the register for each bit that exists within the register, I end up with a hard time remembering all of the bits used within a project. Because of the nature of my work, I have some larger projects that may contain as many as 100 different flags or more. It becomes much easier just to remember the bit name rather than remembering what register that bit resides in. The luxury of having the single bit name already working on some of my first projects drove me to figure out how to make it work with the Cypress assembler.

If you desire to set your variables to known addresses, there are a couple of methods that you can use. Regardless of which method that you select, you should always do a quick check of the mapping file so as to assure that the RAM is not double defined.

Method 1 – Use Equates to Assign RAM to an Unused Area

This is the simplest method. Since you have a good idea where the RAM is being used and how much space that the stack has allotted, you can simply equate your variables to static addresses that reside in the unused area. Let's say that your project is using six variables for user module variables and has allotted the 16 bytes for the Stack_Zone. That would theoretically place the first free byte at address 23 (decimal) or 17h. Note that this allows for the usage of __r0 during boot.asm. You could start your equates for variable usage starting at that address or higher. If you don't need to use all the RAM, then it would be good practice to leave a bit of a buffer before starting your equate statements. This not only builds some more confidence that this address won't be pushed if you need to add some more user modules, but it also will build a bit of room for other projects to grow that need to have different user module configurations that may need to have more growing room.

```
variable1:    EQU    30h
variable2:    EQU    31h
variable3:    EQU    32h
```

The use of the equate statement rather than the BLK statement will allow you to use the labels in your instructions, for example, MOV A,[variable1], but it will not reserve that memory the same way that the BLK statement will. Since method 1 does nothing more than set up the equate statements, it has the potential of leading to disaster. An example that I mentioned earlier is how the 25xxx/26xxx parts need to have the stack placed at address 40h or later if a UART receiver is present. The way this is handled in the boot.asm file will not show the Stack_Zone being placed at 40h. It will show the Stack_Zone residing at a lower address, but will then place the stack pointer to point at location 40h assuming that this is an unused area. Currently, I don't know of any other situations that do this, but if such a scenario is found and the generated files place the stack pointer in a higher address, you could have some unexpected behavior with your project.

The equate statement should also not be defined globally. That is, you shouldn't use the double colon or the export directive with an equated item. Cypress warns that the compiler will not handle it correctly all the time. Therefore, the equate statements should reside in an include file such as ram.inc. The include file can then be included in all the assembly files that need to reference those names.

Method 2 – Use a BLK Statement to Reserve a Chunk of Memory Before the Stack

This method will mix use of the BLK directive with the use of the EQU directive. The BLK directive will reserve a large chunk of memory and the EQU statements will then allot memory within that chunk to the various variables. An example is listed as follows:

```
RAM.asm file -
;This file will contain the allocation for a chunk of memory
AREA        bss(RAM)
GENERAL_PURPOSE_RAM::    BLK        C0h
```

I added the double colon for the label GENERAL_PURPOSE_RAM so that it would show up in the mapping file. It's not absolutely necessary, since it will not be referred to outside this file, although it might be useful as a label for a routine to initialize all the RAM values. The GENERAL_PURPOSE_RAM chunk has now allocated C0h bytes worth of RAM as shown in the mapping file.

```
Area                     Addr   Size   Decimal Bytes (Attributes)
----------------------   ----   ----   ------- ----- ------------
                   bss   0001   00C0 =    192. bytes (rel,con,ram)

        Addr   Global Symbol
        -----  ------------------------------
        0001   __bss_start
        0001   GENERAL_PURPOSE_RAM
        00C1   __bss_end

Area                     Addr   Size   Decimal Bytes (Attributes)
----------------------   ----   ----   ------- ----- ------------
             Stack_Zone  00C1   0010 =     16. bytes (rel,con,ram)

        Addr   Global Symbol
        -----  ------------------------------
        00C1   _stack_start
        00D1   __Stack_Zone_end
```

You can see that the stack now resides at the end of the bss area and there is a large area from 01h to C1h that is currently not used. Now, my ram.inc file will look similar to the example in method 1.

```
ram.inc file -
variable1:   EQU    10h
variable2:   EQU    11h
variable3:   EQU    12h
```

I have still left some unused room before starting my variable addresses, since I may want to add some user modules that will move the starting location of the bss area. How much room to leave is really up to you and your memory requirements. With this method, however, you have told the assembler that you are using the section defined by GENERAL_PURPOSE_RAM for your use and it has placed the stack after that area. In the case where the stack was moved for the receiver module, your project would still run properly, because your memory usage is before the stack.

The only disadvantage that I can see to this method is that as your RAM requirements for user modules change, you may be forced to move the location of the variables that you wanted to be static. This should only present a problem if you are using the large majority of RAM available on your part, however.

Method 3 – Use a BLK Statement to Reserve a Chunk of Memory After the Stack

This method is very similar to method 2, but the memory chunk resides after the stack rather than before it. This is accomplished by declaring a new area of RAM and giving that area a specified address using the ORG directive.

Therefore, your new ram.asm file will look something like this:

```
ram.asm file -
;This file will contain the allocation for a chunk of memory
AREA   bss1(RAM,ABS)

       ORG    30h
GENERAL_PURPOSE_RAM::    BLK       C0h
```

Notice that the AREA directive now has declared a different name, bss1, for my memory. It has also added the key word ABS to declare the bss1 area to be at an absolute address. The following ORG statement specifies that address to be 30h.

The BLK statement will then place the chunk of memory starting at 30h as reflected in the mapping file.

```
Area                          Addr   Size   Decimal Bytes (Attributes)
----------------------        ----   ----   ------- ----- ------------
                   bss1       0000   00F0 =    240. bytes (abs,ovr,ram)

      Addr   Global Symbol
      -----  -------------------------------
      0030   GENERAL_PURPOSE_RAM

Area                          Addr   Size   Decimal Bytes (Attributes)
----------------------        ----   ----   ------- ----- ------------
             Stack_Zone       0001   0010 =     16. bytes (rel,con,ram)

      Addr   Global Symbol
      -----  -------------------------------
      0001   _stack_start
      0011   __Stack_Zone_end
```

A closer look at the mapping file shows that the areas probably aren't handled like you would expect. The bss1 area is not starting at location 30h, but it starts at 00h and goes to F0h. The stack seems to be in the correct location, but the bss1 area looks like it only worked halfway. Using the additional keyword with the AREA directive to concatenate the areas doesn't have any effect either. The mapping file will show the area as being an overwrite anyway.

I would not recommend use of this method. As you move around the area of GENERAL_PURPOSE_RAM and change its size, you will not get errors when you are heading for trouble, such as overlaying the stack zone or running off the end of available RAM. To me this is a little more dangerous than the first method, because you would expect PSoC Designer to flag an apparent problem, when you've been specific about where you want things placed.

CHAPTER 9

Multiple Configurations

Chapter 8, *PSoC Memory Management*, explained a little about how the base configuration of the part is handled by creating files and routines that would load the control registers for the digital and analog blocks to configure your part as selected in the Device Editor pinout and configuration view. This chapter will discuss what happens when you choose to add multiple configurations.

Multiple configurations give you a visual method to make your PSoC alter its operation and configuration of the digital and analog blocks while running. This is intended to give you greater flexibility and power in your designs. It works particularly well if there are functions in your design that don't need to be happening all the time, but only need to happen at specific moments. After a specific task is done, that function can be freed up to perform another purpose.

My first exposure to looking at a multiple configuration setup was exploring the 300 baud modem project. There is a copy of this project on the companion CD-ROM. In a modem function, there are specific times when you need certain capabilities. Multiple configurations were used to enable and disable these capabilities. An example is with the DTMFGen configuration. This configuration includes the resources necessary to generate the dual-tone multiple-frequency (DTMF) tones required to dial a phone line. The configuration is loaded when the modem is given the command to dial. The configuration is then unloaded after the dialing is complete, since it will not be used again until dialing is necessary. This allows the blocks used in that configuration to be freed up for other uses at other times.

In order to illustrate how multiple configurations are handled, this chapter will walk through some simple configurations. This project is rather simple and could easily fit into one configuration, but I want to keep the project simple enough to understand the basics behind how the configurations are handled. I will not be adding anymore code than what is generated by PSoC Designer. I will be using the 27443 part for my example project. If you haven't already completed a project from start to finish, I suggest that you read through Appendix B – *Project Walkthrough* before continuing.

After starting a new project, which I've chosen to call 'multipleconfig', I have selected for modules an 8-bit timer, a 6-bit SAR ADC, and a PGA using the default names for the modules. I placed the modules in the first positions suggested by PSoC Designer. This puts TIMER8_1 in block DBB00, PGA_1 in ACB00, and SAR6_1 in ASC10. I've set port 0 bit 1 to be my analog input for PGA_1. I've also connected the Timer8_1 terminal count output to port 0 bit 4. Finally, I've changed port 1 bit 7 to be a strong drive standard CPU output.

After placing the modules, I am going to select **Loadable Configuration → New** from the **Config** menu. This places a new tab under the toolbars next to the tab that bears the project name. The name of this newly created tab defaults to **Config1**. I will leave the default name for the tab for this example, but you can easily rename the configuration by clicking in the tab and modifying the name. Notice that when you have the **Config1** tab selected that there are currently no modules shown in that configuration. If you click on the default configuration, which is the tab that has your project name on it, then you see that the modules you selected for that configuration are still there.

I am going to select three modules to reside in my new configuration **Config1**. I've selected an 8-bit PWM, another PGA, and a 6-bit DAC. I've placed the module PWM8_1 in the DBB01 block, PGA_2 in the ACB00 block, and DAC6_1 in the ASC10 block. I've now connected port 0 bit 3 to be the input of the PGA_2. I've changed port 0 bit 0 to be the default setting. I've also connected the Compare Output of PWM8_1 to be connected to port 2 bit 6. I've left port 0 bit 4 alone as the global output that was set up in the default configuration.

You may notice that the PGA has the default name of PGA_2. This is because PSoC Designer recognizes that another PGA already exists with the name PGA_1.

It creates a conflict to have multiple modules with the same name, as their routine names and register references need to have unique names. Even though they reside in unique configurations, these routines still have to be referenced in the same code space. PSoC Designer will prevent you from naming multiple blocks with the same name inadvertently. You can test this by trying to rename the PGA_2 block to PGA_1. The name will revert back to PGA_2 upon pressing the **Enter** key, and a note at the bottom of the window will show that you tried to use a name that is already in use.

Finally, let's create a third configuration with the default name Config2. This configuration will only have an 8-bit timer in it. I've left the name at the default Timer8_2 and have placed this timer in the DCB02 block. I've connected its terminal output to port 0 bit 3. After placing this block, switch to the Application Editor and generate the application. After the application has generated, look at the files that have been created in the project view. The files from my project are shown here in Figure 9-1.

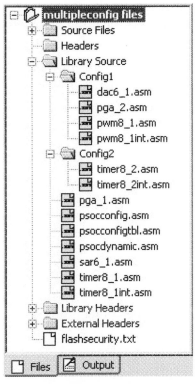

Figure 9-1: Project Tree

PSoC Designer has created subfolders within the library source with the names of the extra configurations. The associated modules for those configurations have their files within the subfolders. This was a nice touch that PSoC Designer adds to help separate the files. If you examine the files as they are stored on your hard drive, you will see that the subfolders Config1 and Config2 do not exist. The file structure that is shown to you here is stored in the PSoCConfig.xml file within your project. The source files for the modules which exist in the default configuration of multipleconfig reside in the library source section and are not listed in a subfolder. The library source also contains the psocconfig.asm and psocconfigtbl.asm as you have seen in your other projects, but it adds a new file called psocdynamic.asm.

I have created another project, called singlconfig, that contains the modules that we have placed in the default configuration of multipleconfig. This will serve as a comparison on how the files will look from a project with a single configuration compared to a project with multiple configurations. These two projects are available on the companion CD-ROM.

Let's start examining the differences by opening the psocconfig.asm file. Near the top of the PSoC config file, PSoC Designer has generated several export statements. These statements allow the other files to recognize labels of subroutines within that file. The list of subroutines available in the multipleconfig project is much larger. My singlconfig project lists the routines that were discussed somewhat in Chapter 8 – *PSoC Memory Management*.

List of export statements from singlconfig project:
```
export  LoadConfigInit
export  _LoadConfigInit
export  LoadConfig_singleconfig
export  _LoadConfig_singleconfig
export  NO_SHADOW
export  _NO_SHADOW
```

List of export statements from multipleconfig project:
```
export  LoadConfigInit
export  _LoadConfigInit
export  LoadConfig_multipleconfig
export  _LoadConfig_multipleconfig
```

```
export UnloadConfig_multipleconfig
export _UnloadConfig_multipleconfig
export ReloadConfig_multipleconfig
export _ReloadConfig_multipleconfig
export LoadConfig_Config1
export _LoadConfig_Config1
export UnloadConfig_Config1
export _UnloadConfig_Config1
export LoadConfig_Config2
export _LoadConfig_Config2
export UnloadConfig_Config2
export _UnloadConfig_Config2
export UnloadConfig_Total
export _UnloadConfig_Total
export ACTIVE_CONFIG_STATUS
export NO_SHADOW
export _NO_SHADOW
```

The best way to understand how to manipulate the configurations is to understand what each of these routines is doing. Notice that there are two labels for each routine. One is used for assembly calls, and one is for C language use. However, both calls go to the same location. It is worth the effort to spend some time examining the routines closely yourself. I will summarize the effects of the routines that are listed here in the order that they have been declared. The declarations are listed in the order that the configurations appear within the project. These routines also appear in this order in the file. However, it appears that despite the declarations, PSoC Designer places the tables in alphabetical order according to the label name in the psocconfigtbl.asm file. This might be a little confusing at first, but **Ctrl** + **F** can bail you out if it gets too bad.

The projects singleconfig and multipleconfig were set to use the loop type of configuration, which is a little more size efficient and is the default setting for code generation. During my summary, some references refer to the project name. Since I'm looking at two projects in this chapter, I will insert *ProjectName* in places where both projects apply. If this usage only applies to one of the projects, I will use that project's name as shown in the code.

Several of the routines listed previously call the LoadConfig routine. This routine will retrieve register addresses and values to load at those addresses. These tables are found in the psocconfigtbl.asm file. A value of FFh signifies the end of the table.

LoadConfigInit

This routine is called in both projects. The LoadConfigInit routine simply calls two other routines before exiting. The other two routines called by LoadConfigInit are LoadConfigTBL_*ProjectName*_Ordered and LoadConfig_ *ProjectName*. The first routine, LoadConfigTBL_*ProjectName*_Ordered, is found in the psocconfigtbl.asm file. This routine will configure all the port registers for the project. These registers for each port on the device include the drive mode registers, the global select register, the interrupt control registers, and the interrupt enable register. You will notice that the port registers reflect the proper settings for the I/O selections of the default configuration.

LoadConfig_ *ProjectName*

The second routine, LoadConfig_ *ProjectName*, will load all the other registers that associate with the CPU core and interconnects. The bottom of this table loads all the configuration registers for blocks used in the default configuration, such as ACB00, ASC10, and DBB00. There is one variance in this routine from the multipleconfig project compared to the singlconfig project. The multipleconfig project sets an active status bit at the end of the routine. This is done with an OR instruction as follows:

```
or [ACTIVE_CONFIG_STATUS+multipleconfig_ADDR_OFF], multipleconfig_BIT
```

The ACTIVE_CONFIG_STATUS byte is declared in the NO_SHADOW area. It will serve as an indicator for what configurations are currently loaded. A quick search for instances of the multipleconfig_ADDR_OFF (multiple config address offset) reveals that this constant is a 0. Likewise the multipleconfig_BIT is a 1 in this project, thereby setting bit 0. The other configurations will set other bits with this register. If more registers are needed the address offset indexes to those registers accordingly.

UnloadConfig_multipleconfig

The tables associated with this routine are as follows:

```
UnloadConfigTBL_multipleconfig_Bank0:
;   Instance name PGA_1, User Module PGA
```

```
;      Instance name PGA_1, Block Name GAIN(ACB00)
        db    73h, 00h    ;PGA_1_CR2 (ACB00CR2)
;   Instance name SAR6_1, User Module SAR6
;      Instance name SAR6_1, Block Name SA(ASC10)
        db    83h, 00h    ;SAR6_1_CR3 (ASC10CR3)
;   Instance name Timer8_1, User Module Timer8
;      Instance name Timer8_1, Block Name TIMER8(DBB00)
        db    23h, 00h    ;Timer8_1_CONTROL_0 (DBB00CR0)
        db    ffh
UnloadConfigTBL_multipleconfig_Bank1:
;   Instance name PGA_1, User Module PGA
;      Instance name PGA_1, Block Name GAIN(ACB00)
;   Instance name SAR6_1, User Module SAR6
;      Instance name SAR6_1, Block Name SA(ASC10)
;   Instance name Timer8_1, User Module Timer8
;      Instance name Timer8_1, Block Name TIMER8(DBB00)
        db    20h, 00h    ;Timer8_1_DIG_BasicFunction (DBB00FN)
        db    21h, 00h    ;Timer8_1_DIG_Input (DBB00IN)
        db    22h, 00h    ;Timer8_1_DIG_Output (DBB00OU)
        db    ffh


;   Instance name PGA_1, User Module PGA
;      Instance name PGA_1, Block Name GAIN(ACB00)
;   Instance name SAR6_1, User Module SAR6
;      Instance name SAR6_1, Block Name SA(ASC10)
;   Instance name Timer8_1, User Module Timer8
;      Instance name Timer8_1, Block Name TIMER8(DBB00)
        db    ffh
```

The unload configuration routine for the default configuration does not affect the CPU registers that were found in the load table. This leaves the CPU and interconnects running as they were before this routine was called. The surprise is that the registers belonging to the blocks used in this routine are not completely returned back to their power-up value of 00. For example, the table area associated with ACB00 which is the PGA_1 block has four different registers that are configured with the load routine. The unload routine only changes one of those four registers. However, only one register needs to be affected to disable this block. The lowest 2 bits of the ACB00CR2 register is reset to 00 remove power from the block. Since power is gone, Cypress must feel that it is superfluous to reset the other registers. If another configuration is going to use the block, then it will reload all four registers, so you shouldn't see a problem in most applications. I suppose that most of you noticed in your casual observance of the tables that the lowest two bits of the ACB00CR2 register were already 00 as dictated in the load config table. However, you must remember that when you call the start routine for the PGA_1 module, those two bits will be affected with the power setting that you dictate. The ASC10CR3 register is cleared, which will disable power to my SAR module, and clearing the DB00CR0 register will clear the enable bit for my timer. The bank 1 table also clears the function, input, and output registers of the digital block. This will disable any connection to the broadcast net, put all inputs into the block into a low state, and disable the primary and auxiliary outputs of the digital block. The last step of the unload routine clears the active status bit for multipleconfig configuration. This routine does not appear in the singlconfig project since it only has one configuration.

ReloadConfig_multipleconfig

The ReloadConfig_multipleconfig routine will simply reload all the registers associated with the analog and digital blocks of the multipleconfig configuration. This table copies the bottom portion of the LoadConfig_multipleconfig. This routine does not appear in the singleconfig project since it only has one configuration.

LoadConfig_Config1

The LoadConfig_Config1 routine is responsible for switching the blocks used in that configuration into a state where those modules can be used. If there are resources needed for this configuration that are being used by a different configuration, the expected process is to have that configuration unloaded before call this load configuration routine. That will take knowledge on the programmer's part about what blocks are used. This also means that if you are only overlapping one resource, then you may not have built your configurations in the most useful way. The Config1 configuration does indeed use the same analog blocks as the default configuration, multipleconfig, but the digital module PWM8_1 is in a different block not used by multipleconfig. When I call the unload routine for the multipleconfig configuration, I will be disabling the Timer8_1 module when I really don't have to. This is acceptable for my project, so I will leave the configurations as they are.

The first step for the LoadConfig_Config1 routine is to set up needed port connections and other CPU registers that need to be changed for this configuration. Let's look at the lines that PSoC Designer inserted here.

```
        M8C_SetBank1
  ; writing Port_2_DriveMode_0 register
        and             reg[ 8h], ~40h
        or              reg[ 8h],  40h
  ; writing Port_2_DriveMode_1 register
        and             reg[ 9h], ~40h
  ; writing Port_2_DriveMode_2 register
        M8C_SetBank0
        and             reg[ bh], ~40h
  ; writing Port_2_GlobalSelect register
        and             reg[ ah], ~40h
        or              reg[ ah],  40h
  ; writing AnalogColumnInputSelect register
        and             reg[60h], ~ 3h
        or              reg[60h],   1h
  ; writing Row_0_OutputDrive_0 register
        and             reg[b5h], ~ 2h
  ; writing Row_0_OutputDrive_1 register
        and             reg[b6h], ~ 2h
        or              reg[b6h],   2h
        M8C_SetBank0
```

Port 2 is being used in this configuration as the output of the PWM8_1 module. The drive mode registers for this port are changed using a combination of OR and AND statements. This is meant to affect bit 6 of port 2 that I've chosen for the output and will not affect the other bits of these registers. The last step of changing port 2 is to enable the global bus connection. You should note that the module is not yet configured for this output that is now up and running, therefore you might experience a glitch on this output during configuration.

The next step is to set up the new analog input selection for the PGA_2 block. These lines change the AnalogColumnInputSelect register (denoted as AMX_IN in the data sheet). The combination of the AND and OR statement here will assert the bottom two bits in the fashion desired without affecting the other bits of the register. This selects port 0 bit 3 to be the input for Analog Column Mux 0.

The last steps deal with setting up the interconnects for this configuration. The first instruction clears a bit of the Row_0_OutputDrive_0 register. This bit disconnects the look-up table (LUT) that used to connect row 0 to the GOE[4] (global output even bit 4). This was an output signal in the default configuration. This interconnect row is not used by this configuration, but it is disabled when this configuration is used. The connection of G0[4] to the global bus was not disconnected, but its connection to the row output is. This is depicted in the interconnect view of PSoC Designer. The final steps of asserting bit 1 of the Row_0_OutputDrive 1 register will set up the LUT output of Row_0_Ouput_2 that connects the signal from the PWM8_1 module to the output port 2 bit 6.

After all the port and interconnect values have been loaded, the tables which are referenced in this routine will load all the needed registers to make sure that the DAC6_1, PGA_2, and PWM8_1 modules will all function properly. Therefore it will load all the control registers for the needed analog and digital blocks. It also loads the function, input, and output register for the PWM8_1 block. The final step for the load routine sets the active status bit for this configuration.

UnloadConfig_Config1

The UnloadConfig_Config1 routine starts out with almost identical functions to the UnloadConfig_multipleconfig routine. It will shut down power to the analog blocks used in the configuration and then shut down the digital block used. The inputs and outputs are also shut down for the digital block. There is, however, something more in this unload configuration routine that isn't in the default routine. The added lines are included here.

```
        M8C_SetBank0
; writing AnalogColumnInputSelect register
        and        reg[60h], ~ 3h
; writing Row_0_OutputDrive_0 register
        and        reg[b5h], ~ 2h
        or     reg[b5h],  2h
; writing Row_0_OutputDrive_1 register
        and        reg[b6h], ~ 2h
; writing Port_2_GlobalSelect register
        and        reg[ ah], ~40h
; writing Port_2_DriveMode_2 register
```

```
        and       reg[ bh], ~40h
        or     reg[ bh], 40h
; writing Port_2_DriveMode_1 register
        M8C_SetBank1
        and       reg[ 9h], ~40h
        or       reg[ 9h], 40h
; writing Port_2_DriveMode_0 register
        and       reg[ 8h], ~40h
        M8C_SetBank0
```

These lines are modifying the same registers that were changed during the load configuration routine. These changes are meant to return these registers back to their original state for the default configuration.

LoadConfig_Config2

The LoadConfig_Config2 routine sets up the port registers, interconnect registers and digital block registers to match the configuration config2. The important thing to learn from this instance is that the configuration config2 only had one digital block used. This digital block (DCB02) is not used in the other configurations. This means that this configuration can be loaded at the same time as another configuration without affecting the operation of the other configuration. This is true if the other configurations don't use port 0 bit 3. This pin is used to reflect the output of Timer8_2. If another configuration is using this pin as a general-purpose I/O or as an analog input, then its operation will be affected. The LoadConfig_Config2 routine also sets an active status bit to show that the configuration is loaded.

UnloadConfig_Config2

The UnloadConfig_Config2 routine will turn off the digital block by turning off the enable bit and disabling any input and output connections. It will also return the interconnect state to the default configuration state and clear the active status.

UnloadConfig_Total

The UnloadConfig_Total routine is meant to disable all the digital and analog blocks with the same method that this used by the unload routines for the specific configurations as previously shown. The active status bits for all configurations are also cleared after this routine is called. This routine does not perform any modifications on the CPU registers or the interconnect registers. Therefore, if you call this routine, those respective bus signals have an undetermined state on them.

NO_SHADOW

This is a declaration in the RAM area of the PSoC. In the multiple configuration project, it holds a BLK statement that allocates a byte called ACTIVE_CONFIG_ STATUS. This byte is used by the configuration routines to show what configurations are loaded and which routines are unloaded.

Using multiple configurations can be a powerful tool, but it pays to be well acquainted with what is affected by the generated routines and how they link together. I have tried to use as few configurations as possible in my projects, as I usually work in situations where code space is at a premium and a complete reconfiguration isn't necessary. You always have the option of reconfiguring blocks and interconnects manually. This works very well if you only have a couple of items to change in your configuration.

It isn't an easy task to make a code generator for multiple configurations. It is very difficult to decide what order to use to turn things on an off to work in every situation. The method for loading configurations has evolved over time. This became readily apparent in some work projects when we noticed a glitch on a PWM output every time the chip powered up. What we discovered is that in the latest method of configuration, we had interconnects connected to output pins in a strong drive mode before the block was configured. In our particular project, this was an unacceptable behavior. Our solution around the problem was to configure that I/O pin to be a standard CPU strong drive pin in the visual configuration. I enable the global select option of the pin after I've entered the main file. This way I don't need to worry about glitches as the power up default state for the I/O pin is a low.

Project Pruning

The PSoC differs from other microcontrollers, since you have to configure the digital and analog resources to operate in the way desired. The configuration may take more code space than you are used to. This chapter will suggest a few methods that you can use to optimize the usage of ROM memory and other chip resources so that you can fit more into less while designing with the Cypress PSoC microcontrollers. You should note that even the best designed tools can't always optimize the code as much as you are able to do manually. Don't be discouraged if you can't just select a few options to make your project fit. I often have to employ several tactics together in order to make a tight project fit well in the end. It's also a good idea to see what features of the project are the most important so that you can trim options that are the least important if you've exhausted all other options.

Options Within PSoC Designer

There are various methods built into PSoC. They are the easiest factors to change and, for most projects, they won't cause any design issues. Many times they are simply an option that you select for your project. Each of these settings is project-specific, so you will need to check the desired options for each project. If you are simply copying all the files and folders in your project to another directory for your next version of your project, then the project settings including your optimizations will be preserved. The first setting that you should be aware of is found by clicking **Settings** from the **Project** menu. Select the **ImageCraft Compiler** tab. There are two settings listed in the code compression technologies section. Even though ImageCraft refers to the

C compiler in your project, the compression options here also will save code in your assembly projects.

Condensation

The first option is condensation. This is an option to remove duplicate code. I would recommend that you start with this option turned off until needed as it alters how your code operates. If you do turn it on, take a close look at what effect it has on your code. I used the file compare command in a command prompt window. It did take a sample project of mine from 9562 bytes down to 9537 bytes. In order to examine the effect that it had on my project, I saved a copy of all the files in the output folder of my project and then used the file compare command to look at the two files as follows:

```
C:\TEMP>fc /n output.lst compare\output.lst > compare.txt
```

The file compare\output.lst is the generated file with the condensation option checked. This reflects the compressed code file.

The first bunch of compare statements simply reflects the changed addresses on calls like the call to _main. Notice that the number in parentheses is the line number of that command statement in the source file.

Comparing files output.lst and COMPARE\OUTPUT.LST

```
***** output.lst
  519:   (0420)   lcall _main            ; Call main
  520:          00B5: 7C 16 38 LCALL _main
  521:   (0421)
***** COMPARE\OUTPUT.LST
  519:   (0420)   lcall _main            ; Call main
  520:          00B5: 7C 15 C5 LCALL _main
  521:   (0421)
*****
```

However, upon closer examination further in the file, I noticed the following:

```
***** output.lst
  6636:   clr_alarms:
  6637:          0C52: 5F 58 57 MOV   [88],[87]
  6638:   (0205)   mov          [current_message],0
  6639:          0C55: 55 57 00 MOV   [87],0
  6640:   (0206)   mov          [override_timer],0
```

```
6641:        0C58: 55 59 00 MOV    [89],0
6642:    (0207) clr_alarmflags::
***** COMPARE\OUTPUT.LST
6636:    clr_alarms:
6637:        0C52: 7C 26 20 LCALL 0x2620
6638:    (0205) mov        [current_message],0
6639:    (0206) mov        [override_timer],0
6640:        0C55: 55 59 00 MOV    [89],0
6641:    (0207) clr_alarmflags::
*****
```

Here is an apparent problem; looking on line 6635 of the listing files, I notice the command that should be compiled in line 6637 is as follows:

```
(0204) mov        [last_message],[current_message]
```

This command was interpreted differently in the compressed file to be an LCALL statement, and the next command to load a 0 into the current message is also not compiled at all. The reason is actually found at the target of the LCALL statement. If I look further into the listing file at location 2620, I find that PSoC Designer has created a subroutine that performs the two commands that were apparently missing.

```
2620: 5F 58 57 MOV    [88],[87]
2623: 55 57 00 MOV    [87],0
2626: 7F       RET
```

The LCALL statement takes a total of 3 bytes. The RET statement takes 1 byte. Therefore, for repeated instances of these two commands, I would normally expect a usage of 6 bytes per instance. However, if I repeat this statement more than once, I expect a one time usage of 7 bytes and an instance usage of 3 bytes to perform the LCALL. This could be shortened to 2 bytes if the CALL statement would suffice. My project contained this combination of instructions only twice, but it netted a savings of 2 bytes by turning this instruction into a subroutine rather than executing the code in line. Looking at the end of the listing file around the location 2620, I found fourteen routines that had been created to save code in this way. That is how the condensation option freed up 25 bytes for me in my project. The cost of this option will be a slight increase in required processing time and a possible increase in stack usage for performing the additional calls.

Sublimation

The second option for code compression is the sublimation option, which removes unused API calls. When you place a module into your project, all the API calls that are associated with that module are generated by PSoC Designer. You may not actually be calling any of the APIs, and therefore they are simply unused code taking up space. An example may be that you start a PWM block, but then you leave it operational and never call the STOP routine. PSoC Designer will search out such unused API calls and remove them from your project. Using this option on my project that started at 9562 bytes, this option reduced my project size to 9154 bytes. That's quite a savings. The output window lists what is not being used during the linking process. I've listed two of the 58 unused API calls here for an example.

```
symbol PWM_BUZZ_bReadPulseWidth is unused
symbol Timer8_1_DisableInt is unused
```

Configuration Initialization Type

The fourth option is found under **Project → Settings → Device Editor**. It is the configuration initialization type. There are two types listed. The direct write method gives me the compiled size of 9562. The loop method gives me a size of 9549. Let's look closer at what is happening.

The direct write method generates some routines that will load the configuration registers for the digital and analog blocks with a series of MOV instructions. An example section of my project is included below. The method is very straightforward. It requires 3 bytes of code space for each register loaded, but operates much more quickly than the loop method.

```
LoadConfigTBL_project:
; Ordered Global Register values
    M8C_SetBank0
; Global Register values
    mov   reg[60h], 29h   ; AnalogColumnInputSelect register (AMX_IN)
    mov   reg[64h], 00h   ; AnalogComparatorControl register (CMP_CR)
    mov   reg[63h], 17h   ; AnalogReferenceControl register (ARF_CR)
    mov   reg[65h], 00h   ; AnalogSyncControl register (ASY_CR)
    mov   reg[e6h], 00h   ; DecimatorControl register (DEC_CR)
    mov   reg[02h], 00h   ; Port_0_GlobalSelect register (PRT0GS)
```

```
mov    reg[01h], 00h  ; Port_0_IntEn register (PRT0IE)
mov    reg[06h], 00h  ; Port_1_GlobalSelect register (PRT1GS)
mov    reg[05h], 00h  ; Port_1_IntEn register (PRT1IE)
mov    reg[0ah], c0h  ; Port_2_GlobalSelect register (PRT2GS)
mov    reg[09h], 00h  ; Port_2_IntEn register (PRT2IE)
```

The loop method creates routines that retrieve the values to be loaded from two tables. The LoadConfig_project routine generated below calls the LoadConfig routine twice, once for the values to load in the bank0 registers and once to load the values for the bank1 register.

```
LoadConfig_project:
    push   a
    push   x
    M8C_SetBank0                                    ;switch to bank 0
    mov    a, 0
    asr    a
    mov    A, >LoadConfigTBL_project_Bank0   ;load bank 0 table
    mov    X, <LoadConfigTBL_project_Bank0
    lcall  LoadConfig                               ;load the bank 0 values
    M8C_SetBank1                                    ;set for bank 1
    mov    a, 1
    asr    a
    mov    A, >LoadConfigTBL_project_Bank1   ;load bank 1 table
    mov    X, <LoadConfigTBL_project_Bank1
    lcall  LoadConfig                               ;load the bank 1 values
    M8C_SetBank0                                    ;switch to bank 0
    pop    x
    pop    a
    ret
```

The LoadConfig routine is included in the PSoCConfig.asm file and is not included here due to size. The carry flag holds the distinction of which bank is to be loaded and the A and X registers point to the table from which to load. The LoadConfig routine reads from the table using the ROMX instruction. The first byte of each row of the table is the register address. The second byte of each row of the table is the value to load into that register.

Since each entry of the table is only 2 bytes instead of the 6 bytes required for each table entry in the direct write method, the loop method does have the additional

ROM requirements of the LoadConfig_project and the LoadConfig routines, but this is outweighed by the savings gained. The tables are located in the PSoCConfigTBL. asm file. I've included the first section of the bank0 table for your reference here.

```
LoadConfigTBL_project_Bank0:
;   Global Register values
    db      60h, 29h    ; AnalogColumnInputSelect register (AMX_IN)
    db      64h, 00h    ; AnalogComparatorControl register (CMP_CR)
    db      63h, 17h    ; AnalogReferenceControl register (ARF_CR)
    db      65h, 00h    ; AnalogSyncControl register (ASY_CR)
    db      e6h, 00h    ; DecimatorControl register (DEC_CR)
    db      02h, 00h    ; Port_0_GlobalSelect register (PRT0GS)
    db      01h, 00h    ; Port_0_IntEn register (PRT0IE)
    db      06h, 00h    ; Port_1_GlobalSelect register (PRT1GS)
    db      05h, 00h    ; Port_1_IntEn register (PRT1IE)
    db      0ah, c0h    ; Port_2_GlobalSelect register (PRT2GS)
    db      09h, 00h    ; Port_2_IntEn register (PRT2IE)
```

The linker allots a certain amount of space for boot.asm. The address at which the linker starts to place relocatable code is found by selecting the tab **Linker** after opening settings from the **Project** menu. My example project has the relocatable code start address set to 110 hex. My project setup doesn't require that much space for boot.asm. I can gain some of that space back by allowing the relocatable code to start at an earlier address. This is a step that you should leave until all of your modules have been placed and configured, as this can have some affect on the amount of space available. I would recommend that you don't change this value until code space becomes a problem.

In order to determine what address is acceptable, you will need to examine the listing file or the output hex file. Interpreting the Intel hex file format isn't very difficult, and there are many references available on the Internet that will explain it. However, I will use the listing file method. It is necessary for you to have successfully compiled your project so that a valid listing file is generated. Open boot.asm and use **Ctrl + End** to navigate to the bottom of the file. Scroll up until you see a label next to instructions, above the final directive statements for various areas. Now, open the listing file in the output folder. It will have the extension .lst. Use the **Find** window, (available in the **Edit** menu), to search for the label that you found in boot.asm. My project has the label IDataDone at the end of boot.asm. I've included part of the listing file below for your reference.

```
(0462) IDataDone:
(0463)    ret
   00DE: 7F    RET
   00DF: 30    HALT
   00E0: 30    HALT
   00E1: 30    HALT
   00E2: 30    HALT
   00E3: 30    HALT
```

The value of 30 hex is the HALT instruction. This is programmed into all unused space as a protection against unwanted operation if control should be passed into this area. My project shows the first available byte available as being DF hex. This is the location I can use for my new start address for relocatable code. This change saves me 49 bytes in this project.

If your project uses assembly code only, then you have an option of disabling C language support. This is done by changing the following line in boot.asm.

```
C_LANGUAGE_SUPPORT:      equ 1  ;Set to 0 to optimize for ASM only
```

If you set the C_LANGUAGE_SUPORT to 0 then it will remove some additional code from boot.asm. In my project this moves my first available location for relocatable code from DF hex to B8 hex. It's important to note that the boot.asm file is regenerated every time you use the generate application command. This will change the C_LANGUAGE_SUPPORT back to a value of 1. If you make the same change to the boot.tpl (template) file in the project directory, then the value will remain 0 the next time you generate the application files.

Design Practices

There are some simple design practices that will help you to keep the ROM usage low in the Cypress PSoC. I would like to list a few practices that you can use.

There are several API calls that may not be called more than once in your project. You can pick up a few extra bytes by placing the code from those API calls that you only call once in line with your own code. Here's an example of how that can be done.

I have an 8-bit timer in my project that I use for general-purpose timing. I currently call the Timer8_1_Start routine and the Timer8_1_EnableInt routine during my initialization phase of the project. These routines are not called again. These routines are very simple as they only set bits necessary to perform their respective tasks.

```
Timer8_1_EnableInt:
_Timer8_1_EnableInt:
   M8C_EnableIntMask Timer8_1_INT_REG, Timer8_1_INT_MASK
   ret
Timer8_1_Start:
_Timer8_1_Start:
   or    reg[Timer8_1_CONTROL_REG], CONTROL_REG_START_BIT
   ret
```

For each such subroutine that I call only once, there is an overhead cost of the RET instruction and the CALL instruction. The RET instruction only costs 1 byte. The CALL instruction takes 2 bytes. If it's necessary to use an LCALL instruction, then it will require 3 bytes to call the subroutine. Therefore I can save up to 4 bytes per API call that I put in my initialization section as shown here.

```
;Initialization Section before optimization
   LCALL    Timer8_1_EnableInt           ;Enable interrupt
   LCALL    Timer8_1_Start               ;Start the timer
;Initialization Section after optimization
   M8C_EnableIntMask Timer8_1_INT_REG, Timer8_1_INT_MASK
   or    reg[Timer8_1_CONTROL_REG], CONTROL_REG_START_BIT
```

Multiple configurations can be another area where you can save a few bytes. The methods for reconfiguration of the digital and analog blocks can sometimes be optimized better by you directly loading the configuration registers to modify a block's operation or parameters. Such action should be considered on a case by case basis. There is the possibility of actually writing your own configuration section that utilizes the power on reset values of the configuration registers. This should only be done in extreme cases.

Other Common Practices

There are some other standard design practices that can help to gain some more space in ROM. Some of these practices use the same base principles as have already been shown in this chapter. The condensation option would change sections of code that had been duplicated into a subroutine that was called, thereby saving code space at the cost of execution time. The same method can be applied to the code that you are writing in your project. If code space is a constraint, you should look at making similar code sections subroutines where possible. Look for macros that are called often that can be changed to subroutines.

Sometimes you need to study your whole project to discover some more innovative space saving techniques. Let's examine a hypothetical example. I want to build an odometer for a vehicle. The odometer has a stored number of trip miles and a number of total miles. I have an input which allows the user to clear their trip miles. There is a different input which clears the total miles. The design constraints dictate that whenever I clear the total miles, I likewise need to clear the trip miles. The first method that I may use to accomplish this would have two subroutines to perform the two tasks.

```
ClearTripMiles:
    mov     [trip_miles_lb],0      ;Clear low byte of trip miles
    mov     [trip_miles_hb],0      ;Clear high byte of trip miles
    ret


ClearTotalMiles:
    mov     [total_miles_lb],0     ;Clear low byte of total miles
    mov     [total_miles_hb],0     ;Clear high byte of total miles
    mov     [trip_miles_lb],0      ;Clear low byte of trip miles
    mov     [trip_miles_hb],0      ;Clear high byte of trip miles
    ret
```

I can save some code by restructuring these two routines as shown.

```
ClearTotalMiles:
    mov     [total_miles_lb],0     ;Clear low byte of total miles
    mov     [total_miles_hb],0     ;Clear high byte of total miles


ClearTripMiles:
    mov     [trip_miles_lb],0      ;Clear low byte of trip miles
    mov     [trip_miles_hb],0      ;Clear high byte of trip miles
    ret
```

Now if I call the ClearTotalMiles routine, then the processor falls through into the ClearTripMiles section. If I call the ClearTripMiles routine, the entry point will perform that task only. I am therefore able to save duplicating instructions and am able to eliminate an additional RET instruction.

The most concise code isn't always the best answer, but continual searching for improvements will help you find the balance between size and flexibility. Added time to scrunch the code when you are in no danger of exceeding your ROM limitations and don't foresee a performance problem could easily be considered wasted time. However, some investment into forward thinking and basic planning can avert future disasters as you can smugly announce, "Oh, I already planned for that." PSoC Designer gives you many simple methods to help reduce code space. Utilize these methods first, as they are the easiest to do, but always be on the lookout for ways that you can optimize your system to fit your needs. The best way to deal with the problem is never to let it develop in the first place.

Design Tips

This chapter is meant to be a collection of miscellaneous design tips that you can apply to your projects using the Cypress PSoC. Some of these tips are mentioned in passing in other chapters, but will be restated here. The tips within this chapter will be grouped into various categories with a bold heading to each category.

Working with Data Sheets

If you want to have multiple data sheets open at the same time or be allowed to work on your project and view a data sheet at the same time, you can navigate to the PSoC Designer Standard User Module Directory:

```
(C:\Program Files\Cypress MicroSystems\PSoC Designer\Data\Stdum)
```

Under that directory, you will find a folder for each user module available in the PSoC Designer. In each folder, there is an HTML file that you can open up in your favorite browser. Since they are just web pages, you can open up as many as you like. Since the data sheets are opened up in a different application, you won't have to close the data sheet in order to work on your project.

Opening web pages within your browser does have a drawback, however. You won't have the navigation tabs at the bottom of the screen like you will have inside PSoC Designer, but the scroll button capabilities and the quick find features of browser programs such as Internet Explorer will make finding any reference within the data sheet document simple and quick.

Note: You shouldn't modify any file(s) within the program directory.
These directories also contain the files that PSoC Designer uses to con-
struct your user modules during application generation. If files do get
modified so that the modules aren't working correctly, you will need to
reinstall the latest service pack or PSoC Designer to correct it.

The latest versions of PSoC Designer have added PDF versions of each data sheet.
These PDF versions can be seen by right-clicking from within the data sheet view in
PSoC Designer and choosing **View Data Sheet PDF**. The PDF version gives you
more viewing and searching options and viewing options than you get in Internet
Explorer.

Shortcut Keys and Navigation Within PSoC Designer

I hope that shortcut keys don't rate too high on the geek meter for you. The successful use of shortcut keys can up your performance dramatically. If you aren't used
to using shortcut keys, try using one or two different shortcuts for tasks that you do
repeatedly like building your project, searching in files or switching from one window
to another. Once you get comfortable with this, then concentrate on the next task
that you do the most often to see if there is another shortcut that can help you there.

PSoC Designer follows standard Windows shortcuts. You can navigate through much
of the menu structure of a program by using shortcut key combinations or by using
the **Alt** key in connection with key letters. Once you've started navigating through
a menu structure, then you are able to use the arrow keys to move around. The Enter
key or Space bar key allows you to select different options.

Start by pressing the **Alt** key once while in PSoC Designer. You will see that certain
letters in the menu items will become underlined. Once they are underlined, you can
select a particular menu by pressing the letter that is underlined in the menu that
you want. After pressing the corresponding button, you can continue selecting items
from within the menu by pressing a letter from the next level from within the menu.
If there is a submenu associated with the item, then you will move to that next menu
structure. If the item is the last object in the menu, then it will select that action.

The other option is to start using the arrow keys once you have pressed the **Alt** key. The down arrow key will drop down the menu items and you can navigate through the menu tree. Use the **Enter** key to select an item within a menu.

All Views –

Alt + Space	Brings up an option menu for the program window
Alt + Tab	Cycles through other active programs
	Continue to hold **Alt** and tap **Tab** to shift multiple programs
Alt + Shift + Tab	Cycles through programs in reverse order
Alt + F4	Closes the program window
Ctrl + F5	Starts the Debugger
F1	Help with current item

Application Editor –

Ctrl + C	Copy
Ctrl + Insert	Copy
Ctrl + X	Cut
Shift + Delete	Cut
Ctrl + V	Paste
Shift + Insert	Paste
Double-Click	Select a word
Shift + Double-Click	Select the rest of the word starting at cursor
	Continues to select the next word for successive clicks
Shift + End	Select the rest of the line beginning at the cursor
Shift + Home	Select the first part of the line beginning at the cursor
Shift + Arrow	Selects text from the original cursor position to next position
Ctrl + Shift + Home	Selects text from cursor position to beginning of file
Ctrl + Shift + End	Selects text from cursor position to end of file
Ctrl + A	Selects all text in file
Ctrl +Z	Undo
Ctrl + Y	Redo
Ctrl + F	Find
Ctrl + H	Replace

Alt + E + I	Find in files (Using **Alt** method described previously)
Ctrl + N	New file
Ctrl + O	Open file
Ctrl + S	Save file
Ctrl + P	Print file
Alt + Enter	File properties
Ctrl + F7	Compile
F7	Build
Ctrl + F4	Close current file
Ctrl + Tab	Cycle through open files
Ctrl + Shift + Tab	Cycle in reverse order

Debugger –

F5	Go
F6	Halt
Ctrl + Shift + F5	Reset
F10	Step over
F11	Step into
Shift + F10	Step out
Shift + F11	Step Asm
F8	Watch variables

One Project for Multiple Parts

Be very cautious if you need to program a PSoC with a file that has been generated for a different part. There are a few problems that can arise when you do this. If the original project was designed for a part with fewer resources, then the file won't contain commands to initialize all the resources of a larger part. This could cause the part you are programming not to operate correctly. I have had success programming a smaller part with a file generated for a larger part in the same family. However, this cannot be done reliably if you are changing families of parts.

The cloning feature of PSoC has given me some trouble in the past when cloning to a different family of parts. I haven't seen a problem changing to a different part within the same family. If you are cloning your project to a different family of parts and your cloned project doesn't work, you may have a difficult time figuring out what is going wrong. The suggested alternative to this situation is to create a new project based on the part that you are moving to, reselect and reconfigure all the modules used and then copy over the files of your creation.

Versions

Version control can be done using the cloning method, but I prefer to simply make a copy of the entire project directory. I rename the directory to reflect the version number. The directory name will not change any configuration within the file. However, I strongly recommend that you change the attributes of your old directory to read only. Then you should immediately compile your new version of your project as soon as you open it. The reason for this is that your project files in the new folder still point to the old folder. If you click on a file from the project explorer when working on the newer version, before compiling, it will open the file from the previous folder. This will continue to happen until the project has successfully compiled. After is has successfully compiled, then it will open the files from the correct folder. If you have changed the attributes of the previous folder to read only, then the files will note that they are read only in the title bar.

PSoC Designer Versions

As PSoC Designer versions change, your project can change. Since there is a fair amount of code that is generated by PSoC Designer, your project might have some different characteristics for how it operates. Be patient. It is not an easy task to try to find a way to configure your part that makes everyone happy. There can be some aggravation because your psocconfig.asm and psocconfigtbl.asm files may not compare well if they have changed the order of generation.

Some of the pitfalls I have seen in changing versions include having my analog reference mux reset in the global values. I've seen a change in the default configuration of modules that don't have a parameter specifically set; in other words, the parameter still has the question mark in it. I've seen the order of how the part is configured changed. This causes glitches on power-up and power-down on certain pins. Power-up and power-down is one area I would recommend that you check closely when moving from one version to another. These glitches might also present themselves during configuration calls.

I've also had the code that I place within files generated by PSoC Designer overwritten. You should place your code in the areas listed within these files to prevent your code from being overwritten. There is a backup directory within your project that should have the previous copy of any such file. However, make sure that you keep your old versions around, because I have been clumsy enough to lose the code both in the generated file and in the backup directory.

Saving Space

Chapter 10 – *Project Pruning* listed some effective ways for how you can save space within your project. There are a few more options that you can look into that require deeper examination of what is going on. These methods are meant to be utilized after the project is nearly complete and you don't intend to generate your application files again.

Multiple or Unnecessary Instructions

Sometimes instructions may be performed multiple times. For example, I have seen instructions switching to banks twice in a row. This was present in the multipleconfig project of Chapter 9 – *Multiple Configurations*, but was not mentioned. You can easily remove one of these instructions to save a little space. A couple of examples are included here.

```
_LoadConfig_multipleconfig:
 LoadConfig_multipleconfig:
    push   a
    push   x
    M8C_SetBank1
    M8C_SetBank0
    ;switch to bank 0

;Part of UnloadConfig_multipleconfig
    mov    A, >UnloadConfigTBL_multipleconfig_Bank1 ;unload bank 1 table
    mov    X, <UnloadConfigTBL_multipleconfig_Bank1
    lcall LoadConfig                              ;unload the bank 1 values
    M8C_SetBank0                                  ;switch to bank 0
    M8C_SetBank0
; clear config active bit
```

Overlaying Routines

The load and reload routine for the default configuration of a multiple configuration project repeat much of the same information. You can move the reload label to the appropriate spot within the load table and thereby save that repetition. Since the table ends with an FFh to signify that the table loading should stop, using the table within a table will not affect operation. The multipleconfig table for bank 1 is given below as an example.

```
LoadConfigTBL_multipleconfig_Bank1:
;   Global Register values
    db 61h, 00h ; AnalogClockSelect1 register (CLK_CR1)
    db 69h, 00h ; AnalogClockSelect2 register (CLK_CR2)
    db 60h, 00h ; AnalogColumnClockSelect register (CLK_CR0)
    db 62h, 00h ; AnalogIOControl_0 register (ABF_CR0)
    db 67h, 33h ; AnalogLUTControl0 register (ALT_CR0)
    db 68h, 33h ; AnalogLUTControl1 register (ALT_CR1)
    db 63h, 00h ; AnalogModulatorControl_0 register (AMD_CR0)
    db 66h, 00h ; AnalogModulatorControl_1 register (AMD_CR1)
    db d1h, 00h ; GlobalDigitalInterconnect_Drive_Even_Input
        register (GDI_E_IN)
    db d3h, 00h ; GlobalDigitalInterconnect_Drive_Even_Output
        register (GDI_E_OU)
    db d0h, 00h ; GlobalDigitalInterconnect_Drive_Odd_Input
        register (GDI_O_IN)
    db d2h, 00h ; GlobalDigitalInterconnect_Drive_Odd_Output
        register (GDI_O_OU)
    db e1h, 00h ; OscillatorControl_1 register (OSC_CR1)
    db e2h, 00h ; OscillatorControl_2 register (OSC_CR2)
    db dfh, ffh ; OscillatorControl_3 register (OSC_CR3)
    db deh, 00h ; OscillatorControl_4 register (OSC_CR4)
    db e3h, 87h ; VoltageMonitorControl register (VLT_CR)
ReloadConfigTBL_multipleconfig_Bank1:
;   Instance name PGA_1, User Module PGA
;       Instance name PGA_1, Block Name GAIN(ACB00)
;   Instance name SAR6_1, User Module SAR6
;       Instance name SAR6_1, Block Name SA(ASC10)
;   Instance name Timer8_1, User Module Timer8
;       Instance name Timer8_1, Block Name TIMER8(DBB00)
    db 20h, 20h ;Timer8_1_FUNC_REG(DBB00FN)
    db 21h, 01h ;Timer8_1_INPUT_REG(DBB00IN)
```

```
        db 22h, 44h ;Timer8_1_OUTPUT_REG(DBB00OU)
        db ffh
```

Relying on Initial Register Values

Many of the register values initialize to 00 on power up. The configuration files will often load 00 out to these registers to ensure proper configuration in event that another instruction has changed them. If you feel confident that you can rely on the power up setting and that you don't need to reassert those register values, you can remove the loading of 00 in many of the configuration registers.

Reducing Bank Switching

Some areas of the configuration files, as they stand, switch back and forth between register banks repeatedly. If you consolidate the entire bank0 register loading into one spot and the entire bank1 register loading into one spot, you can save the space required by switching back and forth. I've included an area from the multipleconfig project that shows an example of where space could be saved.

```
    LoadConfigTBL_multipleconfig_Ordered:
    ;  Ordered Global Register values
        M8C_SetBank1
        mov    reg[00h], 10h  ; Port_0_DriveMode_0 register (PRT0DM0)
        mov    reg[01h], efh  ; Port_0_DriveMode_1 register (PRT0DM1)
        M8C_SetBank0
        mov    reg[03h], efh  ; Port_0_DriveMode_2 register (PRT0DM2)
        mov    reg[02h], 10h  ; Port_0_GlobalSelect register (PRT0GS)
        M8C_SetBank1
        mov    reg[02h], 00h  ; Port_0_IntCtrl_0 register (PRT0IC0)
        mov    reg[03h], 00h  ; Port_0_IntCtrl_1 register (PRT0IC1)
        M8C_SetBank0
        mov    reg[01h], 00h  ; Port_0_IntEn register (PRT0IE)
```

Boot.asm File

Changes to the boot.asm file may become necessary for you to maintain in a project. An example might be where you want to set an interrupt vector for an analog column interrupt. There isn't a visual way to select a location for that interrupt. Every time the generate application files action is used, the boot.asm file is regenerated. It can become cumbersome to redo the same changes over and over again.

You can overcome this frustration by modifying the boot template file boot.tpl. This file is found in your project directory. It is used as the model from which to generate the boot.asm file. Changes in this file will be reflected in the boot.asm file when you generate the application files. I personally don't overwrite any of the code I replace, but I simply mode it out into a comment area so I can restore the original code if needed. An example of how to set up an interrupt vector for an analog column interrupt is shown here.

```
boot.tpl file lines before modification.
    org 08h      ;Analog Column 0 Interrupt Vector
    `@INTERRUPT_2`
    reti

boot.tpl file lines after modification
    org 08h      ;Analog Column 0 Interrupt Vector
    ljmp  my_interrupt    ;`@INTERRUPT_2`
    reti

boot.asm lines before modification
    org 08h      ;Analog Column 0 Interrupt Vector
    // call  void_handler
    reti

boot.asm lines after modification
    org 08h      ;Analog Column 0 Interrupt Vector
    ljmp  my_interrupt    ;// call void_handler
    reti
```

Temporary Removal of Routines

There are various reasons why you may want to disable a section of code temporarily. It might be to minimize complexity of operation to find a problem, or to remove a less important routine temporarily while you work on reducing code sections to save space. PSoC Designer has a comment block button that assists by commenting highlighted areas of text. The comment and uncomment block buttons are shown in Figure 11-1.

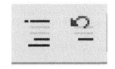

Figure 11-1: Comment and Uncomment Block Icons

When you are commenting blocks of code, make sure that you don't comment the routine labels or the RET instruction at the end of routines. This will save you the trouble of removing references to the routine from other areas of code. Making sure that the RET instruction remains will ensure that calls to the routine won't start executing the next section of code after the routine and thereby produce unexpected results.

Control Systems

Control systems require particular caution in making sure that the design always acts in a responsible and expected manner in all conceivable conditions. The goal is to make sure that the system being controlled is never directed to take an action that will destroy the system or cause danger to surrounding systems, objects, or people. Having done some design on controllers for treadmills gives me some personal experience in this area. Here are some tips that might be useful in your control system.

Switch Mode Pump

The switch mode pump can be very useful to allow your circuit to run off low voltage levels. However, the switch mode pump can also be the source of significant noise. I would recommend not using the switch mode pump to power a system where such noise levels could disrupt operation.

The switch mode pump circuitry will respond faster than the low voltage detection circuitry. If fast power glitches could possibly disrupt operation and not be detected by the low voltage detection circuitry, then you might consider adding a circuit that uses transitions of the switch mode pump to alert the processor of the power dip so it can respond appropriately. Since changes in current are responsible for radiated noise, design the circuit to use as little current as possible.

CPU Speed

Higher CPU speed also requires the voltage supply to adhere to tighter specifications. Close scrutiny of the data sheet and a good relationship with your local field applications engineer for Cypress will help you choose which speed is best for your situation.

Toggling an I/O pin on entry and exit to specific areas of code will give you a quick way to determine operation time with an oscilloscope. Such a check gives you confidence in how your system is really running rather than making a guess.

Power Settings

There are tradeoffs in power and accuracy for the power settings of the analog reference multiplexer, analog blocks, and buffer outputs. I recommend that you take some effort to change power setting combinations methodically in small signal applications to make sure that the accuracy of your system is optimized.

Analog-to-Digital Conversion

The Cypress PSoC has different types of analog-to-digital converters, where most microcontrollers have only one. There are different advantages with the different types of converters. Take the time to read about all the types. You might find that you are shortchanging yourself by using the type you are most familiar with.

Analog Column Clocks

It's easy to get focused on one area of the design and forget that what you are changing affects other areas of the design. The analog column clocks are the same for all blocks in that column. Clocks might be changed for reasons such as setting the frequency range of a filter. However, changing this clock might affect the other blocks in unexpected ways. For example, the SAR type of ADC stalls the processor for a time proportional to the clock frequency. The incremental ADC conversion time also relies on this clock signal. Remember to check the effect on all areas of the column when adjusting this clock value.

Startup Code

There is a delay associated with the power up of the PSoC. This delay may be new for some designers. During the configuration time, your outputs and inputs might exhibit some unexpected behavior. Make sure this behavior is acceptable for your design.

Output Buffers

The output buffers can be very useful during your testing stage to output the processor's analog signals or used in conjunction with a DAC to show the state of a digital signal.

Bit Manipulation

Chapter 8 – *PSoC Memory Management* discussed methods that would allow you to set up a static manipulation of RAM. You can use this ability to create bit instructions that only require one argument in the instruction to specify a bit. The method used will be to equate your bit with the register address offset numerically and added to the mask needed to specify a particular bit. Examples of how this is done are shown here.

The equate statements necessary to define the flags could be kept in an include file that would allow multiple files to recognize the definitions.

```
flags:      equ    20h              ;Register used to store various flags
myflag:     equ    (flags*256)+1    ;myflag uses bit 0 of flags register
yourflag:   equ    (flags*256)+2    ;yourflag uses bit 1 of flags register
thisflag:   equ    (flags*256)+4    ;thisflag uses bit 2 of flags register
```

The macro statements will divide the flag number back into its original components of a register address and a mask value.

```
MACRO  btst                 ;Zero asserted to reflect the state of the flag
   tst [@0/256],@0%256   ;Divide isolates register, Modulo isolates bit
ENDM


MACRO  bset                 ;Set flag using an OR instruction
   or  [@0/256],@0%256   ;Register and bit isolated for OR instruction
ENDM


MACRO  bclr                 ;Clear flag using an AND instruction
   and [@0/256],~@0%256  ;~ instruction gets the complement of mask
ENDM
   btst
```

Now your files can utilize the bit instructions using the flag name only to describe the bit. Here is some example code:

```
SomeRoutine:
    btst     myflag    ;see if myflag is set
    jnz      IsSet     ;jump if myflag is set
    call     DoThis    ;DoThis is executed if myflag is clear
    jmp      Done      ;Exit now
IsSet:
    call     DoThat    ;DoThat is executed if myflag is set
Done:
    ret                ;Exit routine
```

CHAPTER 12

PSoC Express

PSoC Express™ is an innovative creation that will bring the flexibility of the PSoC within reach of any programmer. It allows you to configure and program a PSoC for custom applications without ever having to write a line of code or worry about configuration. It has its limitations, of course, but PSoC Express is designed to give programming power to those who have never programmed before.

The basic concept of PSoC Express is to provide a visual drag and drop Windows interface to build an entire system of inputs, outputs, and interconnecting logic and formulas. PSoC Express will then configure and program a Cypress part of your choice with your new program. It even provides you with schematic diagrams of what circuitry is needed to connect to the I/O pins of the PSoC. PSoC Express requires that you have the PSoC Designer installed, and any needed service packs. PSoC Designer is invoked by PSoC Express, and handed the configuration information and code information necessary for PSoC Designer to complete and compile the entire project.

After installation of PSoC Express, you are greeted by the PSoC Express Helper as seen in Figure 12-1.

Figure 12-1: PSoC Express Helper

The PSoC Express Helper will give you step by step instructions on how to start a new project or build a custom version of an existing design in the Express Designs Catalog. The Express Designs Catalog already includes designs such as a fan control, voltage monitor, temperature monitor, front panel, and others.

This chapter will explore some of the features that are currently available in PSoC Express version 1.1 Beta. By the time this book is printed, many other features will most likely be available. I would recommend that you check the web site to get the latest and greatest version. Unfortunately, I won't have the space to do an in depth description of how everything runs within PSoC Express, but it should get you started enough to explore on your own.

Once inside PSoC Express, you will be dragging and dropping four different types of components onto a workspace area. These component types are input, output, valuator, and interface. Once your components are placed, then you will be assigning parameters and equations that will cause some components to be dependent on the state or value of other components. A line interconnecting the components shows this dependence, as seen in Figure 12-2.

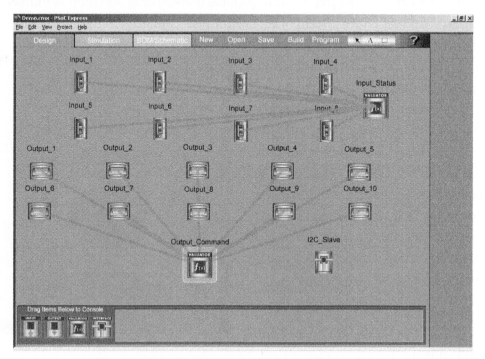

Figure 12-2: Components and Interdependence

The best way to introduce you to PSoC Express is to walk through a project. This project will cover only a few of the capabilities of PSoC Express, but will give you a taste of how powerful the design really is. If you don't have a blank design area already, click on **File → New** to start a new design. It's a good idea to save the project even before starting to prevent warning messages later on. I'll call my project 'Sample'. I'll be starting in the **Design** view that is shown in Figure 12-2. Install the PSoC Express provided on the accompanying CD-ROM and follow along with the chapter.

Design

The first item of our design will be an input. So, start by dragging an input icon from the lower left-hand corner of the design view to any area of your project. Upon releasing the mouse, a window opens that gives you the option to choose what kind of input you want, as seen in Figure 12-3.

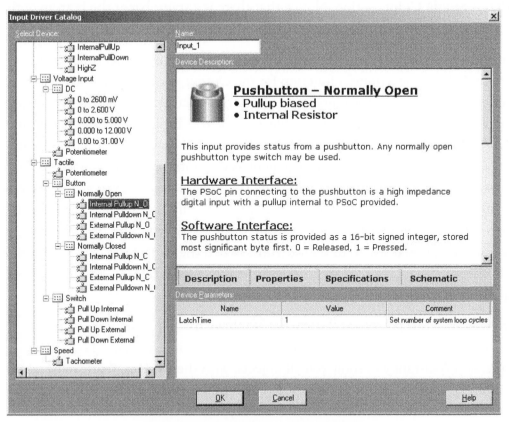

Figure 12-3: PSoC Express Catalog

Figure 12-3 shows how the available input devices are divided into like categories. The familiar tree-like division, similar to the visual representation of folders in Windows Explorer, allows you to find the exact type of input that you desire for your project quickly. I've chosen the **PushButton → Normally Open** object from the Tactile section for this project. You can see that there is a device description area to the right of the device list that shows information about the object that you have selected. The schematic isn't intended to be a schematic of the object. It shows the type of circuitry needed on the outside of the PSoC in order to implement the object. There is a text box at the top center of the window that will allow you to rename the object. I'm going to leave the input with the default name Input_1.

The next step in our project 'Sample' is to select an output. This is done in the same way as you chose the input. Drag the output icon onto the design area and release the mouse button. For this design choose a simple on/off LED from the **Display → LED → Single Color** section. Notice that there is a status bar at the bottom of the window that gives a step by step record of what steps you are doing.

The final option in our design is to make the button, Input_1, affect the LED, Output_2. Entering a transfer function for the Output_2 object will make this connection. Right-click on the Output_1 object and select **Transfer Function**. Three different Transfer functions are available. Choose the **Table Lookup** transfer function. The right side of the screen gives a description of how the transfer function works. This is the same type of arrangement as the truth table that you learned about in basic digital electronics. Once you've pressed the OK button, you are prompted to select the inputs for Output_1. There are only two items listed, since that is all that exists in our project now. The idea of this project is simply to turn on an LED when the button is pressed and to turn off the LED when the button is released. Therefore, only one input state is necessary to determine the output state of Output_1. Select Input_1 and press Next.

Two windows now pop up on the screen. You have the Expression Assistant, as shown in Figure 12-4, and the Add Values window shown in Figure 12-5. The Expression Assistant will show you all objects, with their values, that can be used in your expressions. This is a good list to examine for a while before you select that option not to show this window. If you don't type in the value exactly as it appears here, then the expression will be wrong. Note that there are two underscore characters between the object name and its value.

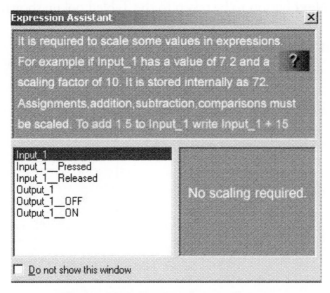

Figure 12-4: Expression Assistant

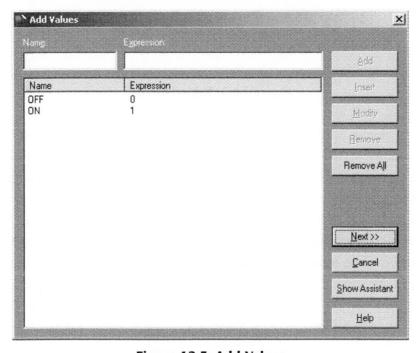

Figure 12-5: Add Values

The Add Value window allows you to add other values to your object output. This project only expects to have the two values for the Output_1 object that already exist. Simply click the **Next** button to be taken to the Table Lookup shown in Figure 12-6.

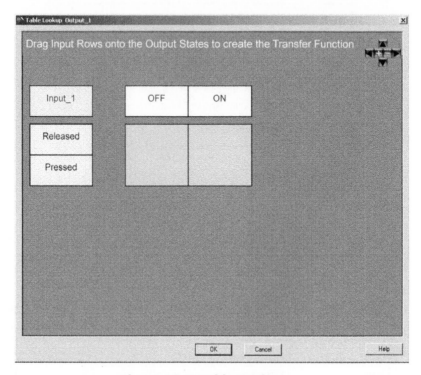

Figure 12-6: Table Lookup

The Table Lookup is our final step to making the correlation between our button input and our LED output. You notice that this table has pan and zoom buttons in the upper right-hand corner. This will help with navigating through much larger tables. For this project, the table is very simple. All that you need to do is to drag a state of Input_1 from the left onto the appropriate state of Output_1 on the right. Once you drop the values, they will align themselves appropriately. For this project, I want the LED to turn on when Input_1 is pressed and off when Input_1 is released. Once you have dropped the inputs into the appropriate columns and pressed the OK button, you will return to the design area and a line will appear showing the dependency that Output_1 now has on Input_1.

Simulation

Click on the **Simulation** tab to start a simulation of the project. If you didn't follow instructions on saving the project earlier, you will be prompted to save it now. The simulation view looks very much like the design view. There will be an addition of a current value area next to Output_1 that will show the current value of Output_1. There is also a current value area that shows up for Input_1. It is to the left of Input_ 1, so if you don't see it you may need to drag Input_1 to the right. There are also controls that look like they came out of a VCR down in the lower left-hand corner of the screen. These controls are used to navigate through a stimulus file that will be recording what actions you are doing during the simulation. These actions, and their generated results, will be recorded to a comma delineated file in your projects directory. There is also a button available to load an externally generated file.

Clicking on the radio buttons of the current value of Input_1, you should be able to toggle the Input_1 state to be pressed or released. The current value of Output_1 will change accordingly. Congratulations, you've just built a PSoC Express Project. Now all that remains is for you to build the project and put it into a PSoC. Click on the word **Build** to continue.

Build

After selecting build from the **Project** menu or by clicking on the word **Build** at the top of your design window, you will be prompted to select a part. The left top side of the window lists the parts available to PSoC Express with a short summary of its resources. The right side of the window lists a more detailed specification of the part. The bottom left side of the window lists any configurable properties that you can change from within PSoC Express. The sample rate determines the loop speed of the system loop cycles that are referenced in some of the objects chosen in the design stage. Press the OK button on the default CY8C24423 part. After the build has been successfully completed you are left with an outline of the part that you have been chosen. Two of the I/O pins of the part now bear the label of our two objects Input_1 and Output_1 as seen in Figure 12-7.

Click on the link near the bottom of the window labeled Sample_BOM.htm. You are taken to a dynamically created HTML file that shows all the required parts to make

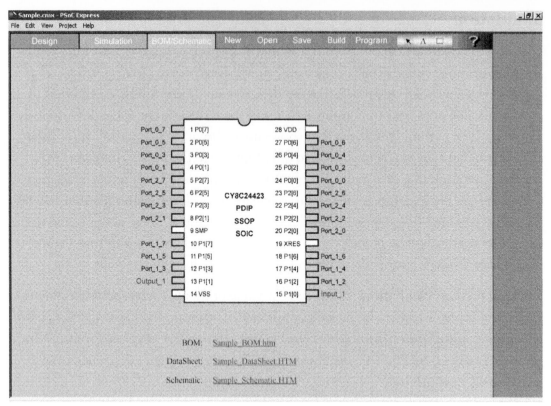

Figure 12-7: Part Outline

your project actually function. This includes labels, descriptions, values and even the Digi-Key part number if known. The actual HTML file isn't in your directory for mailing, however. Hopefully this will happen in the next release.

Click on the link to the sample data sheet and you are taken to a file that shows the part outline along with the pin descriptions in a table format. Each object has a section in this data sheet that lists more information about its particular instance. Information is given about the addressing structure if you should choose to use the I^2C object to communicate with this PSoC part. The storage of object values is also quickly described.

The schematic section again shows the part outline, along with the base device circuit for the part. This circuit is shown here in Figure 12-8.

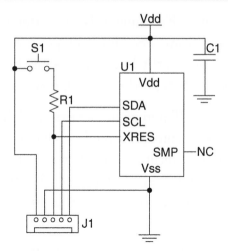

Figure 12-8: Base Device Circuit

This circuit is in addition to the object circuits that you find at the bottom of the schematic window. This circuit provides power and programming capabilities to your part. Programming will be done through the J1 header.

Program

Now that your project has been designed, simulated, and built, you are ready to program your part. After you have built up your circuit, you can click on the word **Program** at the top right of your project or choose **Program** from the **Project** menu and the PSoC Programming application is invoked to program your part. I have been told, however, that even this program may be more complex than what is desired, and it may eventually go away.

Other Transfer Functions

Now that you've grasped the basic idea, let's go back to the design stage and try a different transfer function. Right-click on Output_1 and select the option **Select New Transfer Function**. Since we've already done the Table Lookup option, two options remain, the Priority Encoder and the Status Encoder. The Priority Encoder and Status Encoder are very similar. They will both be constructed using a series of if-then statements.

In the Priority Encoder, the first IF statement that falls true is executed, and the transfer function exits at that point. Therefore, you put the highest priorities near the top. The description shows the Priority Encoder as being the same as an IF..ELSEIF type of statement. Only one THEN portion can happen in the Priority Encoder.

In the Status Encoder, all of the IF statements are executed. If a statement is true, the THEN portion of that line is executed. If several statements are true, all the corresponding THEN statements are executed. This allows you to allow changes that are mutually exclusive, in other words they don't affect each other. If your THEN portions of the transfer function do affect each other, then the last one executed will be the final state. The description aptly likens this to a series of independent IF statements.

For the sake of example, let's redo our project with the Priority Encoder. Select Priority Encoder for the new transfer function. The Expression Assistant pops up again. Move it to the left side for reference. Click in the first IF portion of the transfer function and you will be able to enter the expression for that IF statement. Type in 'Input_1==Input_1__Pressed' (Single quotes are not to be typed but are shown here to frame the needed text). If you entered it correctly then PSoC Express should accept your statement; if you didn't enter it correctly, it will wait patiently. Remember that there are two underscore characters before 'Pressed'. I also have typed two '=' characters in a row that may not look like two characters because of font variations. Now in the THEN section, type a '1'. In the second IF statement type in a '1' for the IF section and a '0' for the THEN section.

The logic here is that if the Input_1 state is currently pressed then I want the Output_1 state to be a 1, which will turn on my LED. If the Input_1 state is not pressed then operation continues onto the second statement which is a 1 and therefore always true. That second statement will always be executed if the first doesn't fall true. The second statement sets the state of Output_1 to a 0, which turns off the LED. Go ahead and press Next to accept the function and test out the simulation, then build the project.

Making a Stimulus File

Currently the stimulus file plays at full speed, so if there isn't something to slow the simulation down, it's hard to tell what is going on. The stimulus file that you are playing is found in your project directory. The file will be named LOG.csv. This type of file can be opened with most spreadsheet type programs. You will need to be out of the simulation view before you can have access to the file, however. The text of a simulation of my current project is included here. The time column shows iterations of the looping structure. A new time value was entered in this file each time I changed the state of Input_1. You can see that the Output_1 state follows the logic of the project. Since you are already in a spreadsheet type program, the response of your system can be easily graphed.

Time	Input_1	Output_1
1	0	0
2	1	1
3	0	0
4	1	1
5	0	0
6	1	1
7	0	0

If you want to create your own stimulus file for playback, I suggest that you copy the stimulus file created in your project file. Then you can modify the Input states using the tools available in your favorite spreadsheet program and open this new stimulus file from PSoC Express.

What is Really Being Done in the Background?

Within your project directory is a subdirectory that contains the PSoC project that is being generated in the background. You will notice in the Device Editor that PSoC Express has created two new modules, SHADOWREGS and the SleepTimer. These modules are best left for use by PSoC Express and shouldn't be used in your normal PSoC projects.

The SleepTimer module controls the normal sleep timer available on the Cypress PSoC. Each time the sleep timer expires, a system loop is performed. The sleep timer therefore is simply used as a delay for execution. This process is more visual if you examine the while loop in main.c. You can see the SleepTimer_SyncWait subroutine which performs this delay called there.

```
while(1)
{
// Sync loop sample rate
SleepTimer_SyncWait(SAMPLE_DIVIDER, SleepTimer_WAIT_RELOAD);

// update input variables
SystemVars.ReadOnlyVars.Input_1 = CMX_PUSHBUTTON_GetValue(ID_In-
    put_1);


// Custom Post Input function
CustomPostInputUpdate();

// run transfer function and update output variables
TransferFunction();

// CustomPreOutputUpdate();
CustomPreOutputUpdate();

// set outputs
CMX_DIO_SetValue(ID_Output_1, SystemVars.ReadOnlyVars.Output_1);
}
```

After the sleep timer provides a delay, the system loop begins by reading the latest state of all the inputs, performing any post processing of the inputs before performing the transfer functions for the project. After that any preprocessing on the outputs is done before asserting the final state of the outputs.

Let's examine the TransferFunction routine. It is found in the transferfunction.c file. This TransferFunction routine only calls one other routine, Output_1_TransferFunction. If we had several transfer functions defined, then each would be called here in turn. Most ordering on the same level is done alphabetically, so name your functions accordingly. If one function depends on another function then the order is altered to fit that requirement.

The Output_1_TransferFunction is found in the systemvar.c file. It reflects the Priority Encoder transfer function that I last specified in my project before I built the project.

```
void Output_1_TransferFunction()
{
        if (SystemVars.ReadOnlyVars.Input_1==Input_1__Pressed)
    {
        SystemVars.ReadOnlyVars.Output_1 = 1;
    }
    else if (1)
    {
        SystemVars.ReadOnlyVars.Output_1 = 0;
    }


}
```

The Valuator and Interface

The valuator and interface are two other objects that we didn't place on our first project. The valuator provides a memory location for you within your project. It has two types, the interface valuator and the transfer function valuator. The interface valuator is a memory location to which you write a value. It is not affected by any inputs other than a direct command to change its value. The transfer function valuator offers you to set up inputs similar to the output object to determine the output value of the valuator. It has the priority encoder, status encoder, and table lookup transfer functions just like our LED did. It also adds a new transfer function called set point region. The set point region divides a range of values into two conditions, values above the set point and values below the set point. You are also given control over a hysteresis value for the setpoint. The set point region transfer function works well for items such as voltage inputs, which could have any value within a given range.

The interface object loads an I²C slave module into your project, which gives your PSoC Express design communication capabilities. Various objects within the design can be addressed for this communication. The interface object allows you to set the address for the device so that your PSoC Express design won't conflict with other I²C devices on the same bus.

The Future of PSoC Express

PSoC Express is going to offer the flexibility of the PSoC with an ease of use never before known. Newer versions will boast more objects to place with more configurability. Right now, any changes that you make to the PSoC project after building in PSoC Express will be overwritten on the next build from PSoC Express, but exceptions for that may be made in future versions. These improvements will have you taking PSoC kits home to your kids to give them the chance to start microcontroller development long before they ever graduate from high school.

APPENDIX **A**

Global Resources

The global resources loaded during the execution of boot.asm are configured in the interconnect / pinout view. These global resources and their function are listed below. This is not intended to be an exhaustive explanation of each resource, but is provided as a quick overview that will allow you to decide which configuration would best suit your project. Make sure to refer to the specific data sheet of the part you are using before setting up the global resources for your project.

The global resources of the original parts that Cypress produced have been carried over to the newer families. However, there have been a couple of additions in the newer families. I will note when a resource is only applicable to the newer parts. Older parts refer to the 25xxx/26xxx families and the newer parts are all other PSoC families.

CPU Clock

The CPU clock specifies the timing of the instruction clock. This global setting asserts bits 0–2 of the OSC_CR0 register in bank 1. The available settings are as follows:

24 MHz	SysClk / 1
12 MHz	SysClk / 2
6 MHz	SysClk / 4
3 MHz	SysClk / 8
1.5 MHz	SysClk / 16
750 KHz	SysClk / 32
185.5 KHz	SysClk / 128
93.75 KHz	SysClk / 256

Should you need to adjust the CPU clock by a direct write to the OSC_CR0 register, then you should take note that the bit values associated with these frequencies don't follow the order that you may expect. Please refer to the data sheet for the proper values.

Each of these clock frequencies is based on a 24 MHz internal main oscillator (IMO). You are able to use the following taps off of the IMO to run your CPU slower, but that does not disallow the usage of the higher frequencies to clock the digital or analog blocks within your project. The other clock sources such as 24V1 and VC1 operate independently of this setting. This allows you to slow down the CPU to save power in various ways without compromising fast digital transactions that need to happen, such as a quick PWM with slow CPU operation.

On the newer parts, the CPU clock frequency has a divider listed next to the Frequency value. This is because the CPU clock of the newer parts is based on SysClk. SysClk is simply the 24 MHz internal oscillator by default, but has the option of being an externally provided signal also. If SysClk is an externally provided signal whose frequency is different from the 24 MHz, then this external frequency divided by the value listed next to the CPU frequency will be the actual value of the CPU frequency.

Slower CPU clocks mean saved power, less noise emissions and greater noise immunity by allowing more time for the signals within the CPU to stabilize. Before setting the oscillator to higher frequencies, verify the accuracy of your power supply and compare that with any limitations listed in the data sheet, to ensure reliable operation. There are also some alignment issues when operating at 24 MHz with the older parts. PSoC Designer will warn you if these conditions exist. There is also an option within your project (when working with older parts) that will allow you to enable PSoC Designer to try to work out the alignment issues automatically. In my experience, it does a good job and only has problems when your code space has very few bytes of code space left.

The internal oscillator of the PSoC has two large factors that affect the accuracy of the internal oscillator; voltage and temperature. The more significant of the two is voltage. If accuracy is an issue, and you intend to use the internal oscillator, pay particular attention to these two factors to ensure the most accurate oscillator possible.

32K_Select

The 32K Select allows you to choose the source of the 32.768 kHz base frequency of your part. It asserts bit 7 of the OSC_CR0 register in bank1.

If you select the internal option, then the 32 kHz will come from the internal main oscillator. If you keep your voltage source and operating temperature in a nominal region, then the oscillator should be good to ±2.5% accuracy. Again, the accuracy of the voltage has the largest effect on the accuracy of the oscillator. The temperature of the die has a lesser effect, but can still swing its accuracy. The internal option also allows you to use the I/O pins P1[0] and P1[1] as general-purpose I/O.

If you select the external option, then the part will use an external 32.768 kHz crystal connected to pins P1[0] and P1[1]. These pins must be set to the High Z or High Z Analog setting to allow the oscillator to work correctly. You will not be able to use the port pins P1[0] and P1[1] as general-purpose I/O any longer. The external crystal should be a much more accurate source for the 32.768 kHz clock source available to the digital and analog blocks inside the processor. Please note however, that the external option does not affect the CPU frequency and SysClk frequency unless used in connection with the PLL_Mode.

PLL_Mode

The PLL mode is a phase locked loop selection option. It controls a phase lock of the internal main oscillator to the external crystal. It asserts bit 6 of the OSC_CR0 register in bank1.

The PLL mode should be set to the **Disable** option, unless you are going to use an external source for the 32 kHz clock. If you want the 24 MHz and other clock taps in the processor to be phase locked to the external frequency, then you select the **Ext Lock** feature.

Since the 32 kHz is unstable at startup, PSoC Designer will add code to the boot. asm file necessary to set up the sleep timer to delay to allow the oscillator to stabilize before switching over to lock to that frequency. Since PSoC Designer takes care of this for you, you don't need to do more than select the option. Be informed, however, of what is going on inside boot.asm so as not to be caught with a problem because of added boot up time.

Sleep_Timer

The sleep timer allows for four settings based on divisions of the 32 kHz clock. It asserts bits 3 and 4 of the OSC_CR0 register of bank 1. The settings are as follows:

512_Hz
64_Hz
8_Hz
1_Hz

The sleep timer is used for both sleep and watchdog functions. The sleep timer sets the period of the micro's sleep time. This sleep interrupt will allow the processor to return from a low-power state to poll system resources to determine if it should be woken up or not. Therefore, if you have a slower sleep timer, then you will use less power because you will be in the low-power sleep state for a longer period of time. Having a faster sleep state will let you respond quicker to a condition that dictates an exit of the sleep condition.

The watchdog period expires on the third cycle of the sleep timer. This will cause your watchdog to expire in between 2 to 3 periods of the sleep timer. The actual time is dependent on the value of the sleep timer when you last refreshed the watchdog timer.

VC1 (24V1=24MHz/N)

VC1 is found on the newer parts. It is a frequency derived from SysClk. Its frequency is SysClk divided by a selected value of 1 to 16. It asserts bits 4 to 7 of the OSC_CR1 register. The divider value is the binary value contained in that nibble, plus one. It can be used as a clock source for both digital and analog blocks.

24V1 is found on the older parts. It is very similar to VC1 and derives its frequency from the 24 MHz internal oscillator.

VC2 (24V2=24V1/N)

VC2 is also found on the newer parts. It is a frequency derived from VC1. Its frequency is VC1 divided by a selected value of 1 to 16. It asserts bits 0 to 3 of the OSC_CR1 register of bank1. The divider value is the binary value contained in that nibble of the OSC_CR1 register plus one. VC2 then becomes a clock source of the frequency of VC1 divided by the divider value. VC2 can be selected as a clock source for both digital and analog blocks.

24V2 is found on the older parts. It is very similar to VC2 and derives its frequency from the 24V1 frequency.

Analog Power

The analog power setting asserts bits 0 to 2 of the ARF_CR register in bank0.

The analog power setting allows you to turn the switch cap blocks on or off and set a low, medium, or high power setting for the reference voltage. The switch cap option needs to be SC On if you are utilizing any function of the switched cap analog blocks. It can be left off to conserve power if you aren't using those blocks. I recommend that you set the power setting to high and then look at reducing the power level only if needed for your project. You should have your reference level set at least as high as the highest power you are using in any of the blocks.

Selecting the All Off option turns off power to all analog on the chip. This is a useful option to use during the sleep period, to allow the processor to reduce power usage. A logical AND of ~30h to the ARF_CR register prior to executing the sleep function will accomplish this. Note that upon exiting the sleep state you will need to restore this register in order for your analog functions to work correctly.

Ref Mux

The ref mux option asserts bits 3–5 of the ARF_CR register of bank 0. This option selects the source of AGND. It also selects the high and low reference signals for analog functions. This is a quick trap for those not familiar with the PSoC.

You might start a quick PSoC project and want to read an analog signal into an analog-to-digital converter. You choose the SAR6 because it's pretty straightforward, but quickly get confused because you aren't reading back the values that you expect.

PSoC Designer defaults having a AGND of Vcc/2. This is probably what you wanted, but it also defaults to a high low reference of ± bandgap. This isn't the full range of GND to Vcc for your project, so your A/D converter is working with a much smaller range. If you select the option Vcc/2 ± Vcc/2, then you will have your analog scaled to the full range of GND to Vcc. The PSoC analog had some limitations in its analog as it approached the rails in this part (see Chapter 4 – *Limitations of the PSoC*), so it was best to design to a smaller range. This would be my guess as to why the part defaults to this option. Note that this range is also used for digital-to-analog converter (DAC) operations. With the smaller range you will also only be able to output to a smaller range on the DAC.

Bandgap is equal to 1.3 volts for this part. If you want you can select 2*Bandgap for your AGND and a high/low reference of ± bandgap. The advantage here is that you have a stable range even though your Vcc may be somewhat less than consistent. This might be attractive for battery powered applications where you can expect a slow drain on the power supply.

There is also an option to allow you to derive these reference levels from P2[4] and P2[6]. P2[4] can be used as the source for the AGND signal and P2[6] can be used as the source for the high/low reference levels. This port pin is only present on the parts that are 28 pins or larger. Note that these signals must conform to the analog level limitations of the part. Refer to the data sheet of the part you are using for these limitations.

The newer PSoC parts offer a couple more options with the ref mux. You can also have Bandgap * 2 ± Bandgap or 1.6 * Bandgap ± 1.6*Bandgap. These two option values are reserved on the older parts.

Op-Amp Bias

The op-amp bias option asserts bit 6 of the ARF_CR register of bank 0. It controls the bias level of all analog functions.

The data sheet tells me that normally this setting should be left at low, but can be set high if a faster slew rate is desired. The faster slew rate, however, also limits the voltage swing slightly and produces higher noise.

I personally haven't changed this level in any of my projects. Unless you need the faster slew rate, I would recommend staying with the Cypress suggestion of leaving this value at its default low setting.

A_Buff_Power

The A_Buff_Power setting asserts bit 0 of the ABF_CR register of bank 1. Leave it at its default low setting unless you are going to use one of the analog output buffers. A high setting for power will push more current into the buffers and make them stronger. On the older parts in particular, pushing signals close to the rails of the analog buffers cause a significant increase in current draw. This can be enough to disrupt small analog signals. I haven't seen the problem with the newer parts, but I'm not sure what improvements were made in that area.

Switch Mode Pump

The Switch Mode Pump option asserts bit 7 of the VLT_CR register of bank1. It allows you to turn the switch mode pump on or off. The switch mode pump is used to pump an inductor to provide high enough voltage to the PSoC part. It can be used in connection with a single 1.5-volt battery to provide the needed 3.3 volts for the part to run. It can also be used to pump up the voltage in a 5 volt operation setting. The level at which the switch mode pump begins switches is set by the trip voltage setting.

I have used the switch mode pump as a trip signal for my system to signal that a voltage dip occurred since it will trip on voltage disturbances that are of a short enough duration that the low voltage detection does not detect it. This was accomplished by connecting the output of the switch mode pump to an inverter which connects to the reset pin of the PSoC. I would recommend consulting with Cypress before using this method on significant projects, as there are specific considerations to take into account.

Trip Voltage[LVD (SMP)]

The trip voltage option asserts bits 0 to 2 of the VLT_CR register of bank 1. It sets the level at which the low voltage detection occurs and at which the switch mode pump triggers. The PSoC micro has tighter voltage requirements when clocking at faster speeds. You should make sure that the trip voltages are set at a level that ensures consist and accurate operation.

Make sure that you check tolerance values for this setting in the data sheet. The data sheet that I'm referencing for the 27xxx part doesn't show the same voltage settings as are displayed inside PSoC Designer. Hopefully, this will be consistent in future versions of the data sheet.

Supply Voltage

The supply voltage allows you to select 3.3 volt or 5 volt operation. This selection is used to determine which oscillator trim value to retrieve for operation. To my knowledge, this setting isn't actually stored in a register, but is simply used during code generation to determine which parameters need to be retrieved using the supervisory ROM tables.

Watchdog Enable

The watchdog enable option determines if the watchdog function is enabled by default on the execution of boot.asm. The watchdog frequency is related to the sleep frequency selection as previously shown.

Project Walkthrough

Since I believe that example is often the best teacher, I decided to devote a chapter to building a complete PSoC project from scratch. This project will be using a CY8C27443 part and it has the task of blinking an LED at a rate of 0.5 Hz. Each section listed in this chapter will include a description of the process to build the project. This project is built using PSoC Designer 4.1. Care has been taken in this project to allow you to build the same project using a 26xxx/25xxx part. The project will use resources in the blocks and on the processor that are available to both families. However, I would recommend that you get the newer parts if possible as their capabilities are enhanced.

I've included the project files in the examples folder on the accompanying CD-ROM. You can open this project separately while completing the bare bones project below to use as a comparison to the project you are making. The schematic for the project is shown in Figure B-1. It simply provides power to the PSoC and attaches an LED to the I/O pin P0[7].

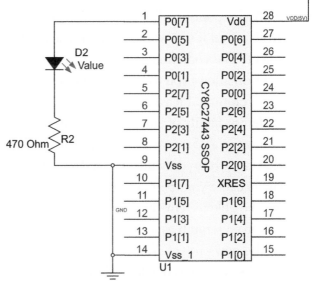

Figure B-1: Barebones Schematic

Setting Up the Project

Upon starting PSoC Designer, a dialog box entitled *Start* opens and allows you to open an existing project or to start a new project. If you are opening an existing project, then you use the **Browse** button to navigate to that project's .SOC file and then select the Device Editor, Application Editor, or Debugger. Since we are going to build a new project from scratch, select the **Start New Project** button.

After pressing the **Start New Project** button, PSoC Designer opens the **New Project** dialog box. This allows you to enter a project name. Type the name of your project in the new project name text box. I will use the name bare bones for my project.

You will notice that the new project location text box appends the project name to the folder name. This is a convention of Cypress. If you browse to the location where you want to save your project, then rename your project; the project will be stored in subdirectory of the project name. If you set the location of the project after naming it, then the project will be stored at the location that you select.

To the left of the dialog box, there is a list box titled *Select Method*. This list box allows you to create a new project, clone a project, or create a design-based project. We are building a new project from scratch, so we will select the **Create New Project** option.

The **Clone Project** option allows you to build a different project off an existing project. This is useful if you want to match the analog and digital block layout of an existing project. The code that you've entered into that project is copied over also. The clone option allows you to select a different part than the part selected for the project you are cloning. Please note that it is not recommended to clone a project built on a different family of parts. For example, you shouldn't try to clone a project that was built using a 25xxx or 26xxx part into a 27xxx part. This has caused some problems that have been difficult to find. It has the potential of you scratching your head for days because your part simply doesn't want to work. Future versions of PSoC Designer may be able to remedy this situation better, but until then, plan on creating a new project from scratch as we are doing now and then open your old project side by side with the new project. You can then easily refer to configuration settings and copy over blocks of code as needed. Then you can be sure that you are not fighting a conversion issue when your project isn't working as expected.

The **Create Design-Based Project** option allows you to create a project that is based on an existing design catalog that Cypress includes in the PSoC Designer installation. After selecting the project that you want to start with, you are allowed to set the default configuration of the project. This is the configuration that will be loaded into your part on power-up. At the end of the day, it's very much like the clone project option.

Pressing **Next** will bring up the **Create New Project** dialog box. This box will allow you to select the part that you want to use for your project. The **Family** drop-down list allows you to select which family of parts you want to use. This selection dictates which parts are available in the **Part** drop-down list. The **View Catalog** button will bring up a quick reference chart that will allow you to decide which part will best fit your needs. Please take note of the **Part Image** option within the catalog view, which allows you to examine the form factor of the part including pin assignments.

The final option is to decide whether you want to have your Main file generated using C or assembly. This example will use the assembler option. Unless you have a license for the C compiler, the C option will be grayed out.

The final step is pressing the **Finish** button. This will create the base files needed to build your project. The status bar at the bottom of PSoC Designer describes each stage of this process. Once the needed files have been created, PSoC Designer will open to the **Device Editor → User Module Selection View**.

User Module Selection View

The user module selection view allows you to select what pre-built modules you want to use in your project, view design information about the modules, and monitor device usage at the same time.

The left column of the user module selection view shows what modules are available in PSoC Designer. Clicking the gray group name allows you to view individual modules for selection. Our bare bones project will be quite simple, so all we are going to put in right now is an 8-bit timer. Pressing the Timers group name displays the list of available timer modules for this part. Click on the **Timer8** icon. The Timer8 data sheet will now be displayed at the bottom of the screen. You will notice tabs along the bottom center of PSoC Designer. There are tabs inlaid in the bottom of the data sheet that allow you to jump to specific sections of the data sheet. The data sheet lists the resources needed to implement this module if you are going to use it in your project.

The top center of the window shows the currently selected configuration and which modules exist in that configuration. The center of the window displays a diagram representation of the block that is currently selected. Immediately to the right is a running indicator of the micro's resources, both what is available and what is used.

This bare bones project is going to use an eight bit timer. First I select the 'Timers' group name and then double-click the Timer8 icon. A Timer8 module is then added to the project. You can select and then right-click on the module in order to rename it. It is preferable to rename the modules as desired at this stage to prevent doing more work later on. For this example, I will leave the name of this module as Timer8_1. You will also notice that the resource meter on the right side of PSoC Designer now shows that I have used one of the eight digital blocks and 70 of the 16384 bytes of ROM.

Interconnect View

Select **Interconnect** from the **Config** menu. This will enter the interconnection and pinout view. This view is used to configure global resources, place and configure selected modules, configure interconnections and configure I/O lines.

First we'll familiarize you with the various sub windows of the interconnect view; refer to Figure B-2.

Starting at the top left of the screen, you will see a section titled Global Resources. It allows you to adjust settings that apply to the entire microcontroller and not just one module. The function of each of the global resources and their possible settings are discussed in more depth in Appendix A – *Global Resources*.

Immediately below the Global Resources section is an area titled User Module Parameters. This allows you to adjust the parameters for the currently selected module. Since we don't have a user module placed at the moment, it will be blank. We will address this area after placing our Timer8 module.

Below the User Module Parameter section is an area that allows us to set up the drive of each I/O pin. They will default at a HighZAnalog state that minimizes current

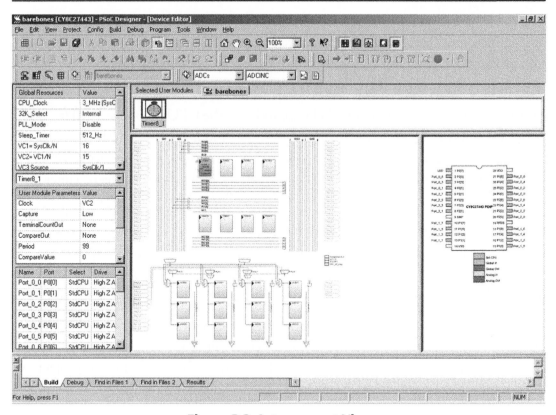

Figure B-2: Interconnect View

into the I/O pins and disconnects the digital reading of each pin. This is to minimize power usage. This area will allow you to assign a name to each pin that is used for identification purposes in the generated data sheet. This name is not currently used as an identifier for the pin within the coding of the project.

In the center of the window, you will see the analog and digital blocks of the part along with graphic depictions of the interconnects between blocks and I/O pins. A pictorial representation of your part is on the far right. It shows pin locations of the physical package and color coding of the default drive state of each pin for the current configuration.

While navigating these two areas, you can hold down on the **Ctrl** key will clicking the left mouse button to zoom in for a closer look. A **Shift** + **Ctrl** click will zoom back out. Holding down on the **Alt** key will let you *grab* the display to pan around while maintaining your zoom.

The top center shows the modules available for the current configuration. If your project has more than one configuration then you will see a tab for each configuration available. In this project, we are only creating one configuration. Its name is the same as the project name, bare bones.

This project is very simple, so I will choose to leave the CPU_Clock at 3 MHz. The 32K_Select and PLL_Mode I will leave at Internal and Disable respectively, because my design doesn't have an external crystal to drive or to sync to. The Sleep_Timer will not be used in this project, so the default setting is okay for the Sleep_Timer also.

The timing taps VC1, VC2 and VC3 need to be set up to allow our timer to provide the 1 Hz necessary to make our project Flash the LED at 0.5 Hz. First set the VC1 value to 16. That will divide the current SysClock of 24 MHz down to 1.5 MHz. Then set VC2 to 15. This takes the VC1 frequency of 1.5 MHz down to 100 kHz. All other global resources will be left at their default values.

VC3 has the capability to divide by a much larger value of 256. Although I could utilize VC3 to join the cascade and divide down closer to my target frequency, I will choose to leave it out of the equation. I do this so that others may use this project in conjunction with the older 25xxx/26xxx parts that don't have the VC3 resource available.

Let's place the Timer8 module in the project. You will notice that the top left digital block of your part has a colored highlight around it. This is the area where PSoC Designer would like to place your part. You can iterate through all possible placements of this block by choosing **Next Allowed Placement** from the **Config** menu or by pressing the corresponding icon in the toolbar. This will move the highlighted outline to the next block in turn and will return back to the first allowed position after all possible locations have been shown. Once you have selected which block you would like to use, then you can select **Place User Module** from the **Config** menu. There is a corresponding icon that will also allow you to place the user module, or you can select **Place** from the **Selection** menu that appears when you right-click on the module in the **Selected User Modules** area.

Please note that even though Designer may show a block as a possible location for a module, it doesn't mean that you can place the module there without causing problems. For example, if I had multiple digital modules selected for this project, I would not be able to assign different modules the same digital block in the same configuration. It would cause a conflict, so PSoC Designer will not place the block when you ask it to. Sometimes the reason may not be clear. You will see a description for the reason the block was not placed on the bottom status bar of PSoC Designer.

I have placed my Timer8_1 module in the top left block designated DBB0. After placing the block, zoom in for a closer look by clicking on the block while holding down on the **Ctrl** key as stated above. Now that I've placed the block, the User Module Parameters area of the screen is filled with parameter names with question marks for values. The question mark doesn't mean that the block will have those values undefined, it just means that they will come up in the default state. I would recommend that you assert all parameters of your blocks so that it will be clear to the next programmer who has to look at your project what values those parameters will have on start-up. There is some advantage to leaving the question marks, in that it could allow PSoC Designer to shorten its initialization section, but I wouldn't worry about that small savings until your project is stretching for the last few bytes of space.

Here are the timer parameters that I will select for my bare bones project.

    Clock = VC2
    Capture = Low
    TerminalCountOut = None
    CompareOut = None
    Period = 99
    Compare Value = 0
    Compare Type = Less Than or Equal
    Interrupt Type = Terminal Count
    Clock Sync = Sync To SysClk
    TC Pulse Width = Full Clock
    Invert Capture = Normal

Some of the parameters listed above are not available on the 25xxx/26xxx parts. These parameters will not be of any consequence on this particular project. I have listed the functions of each parameter as found in the 8-bit timer data sheet below. I've also added some comments in *italics* as to the function of each parameter.

Clock

The clock parameter is selected from one of 15 sources. These sources include the 48 MHz oscillator (5.0 V operation only), 24V1, 24V2, other PSoC blocks, and external inputs routed through global inputs and outputs.

The clock is the heartbeat of the digital block. All state changes of outputs happen on the rising edge of the clock source. The clock source has the option of coming from the 32 kHz source of 15 different sources. It can come from on the internally generated clocks (VC1, VC2, VC3, SysClk*2, etc.) or from external pins. It can also be passed on the global buses from one block to another. The clock can go up to 48 MHz, but is not supported for higher frequencies. Take care to make sure that the clock source is stable before starting the timer, to guarantee desired operation.

Capture

This parameter is selected from one of 16 sources. A rising edge on this input causes the count register to be transferred to the compare register. The software capture mechanism will not operate correctly if this parameter is set to a value of one or is held high externally.

The capture input will allow you to capture the timer value at the event of this input changing states. For the bare bones project, we are not going to capture an event, so I have chosen this value to be low. The low and high options are straight logic values that remain set under all conditions. You will notice that in addition to the global buses that allow you to select this input from other digital blocks and I/O pins that you have the comparator bus options to set your capture. The comparator buses are outputs from the analog block section and allow you to trip the capture event off an analog level.

TerminalCountOut

The terminal count output is an auxiliary counter output. This parameter allows it to be disabled or connected to any of the row output buses. This parameter appears only for members of the CY8C27/24/22xxx family of PSoC devices.

*It seems a little misleading to say that this parameter only appears on the 27/24/22xxx parts, as the same function is available on the 25xxx/26xxx parts, but was listed under a different name **Output**. It provides a single pulse on this output when the timer reaches its terminal* count. You will notice that you are able to route this signal to the global buses to send to an I/O pin or into another digital block.

CompareOut

The compare output may be disabled (without interfering with interrupt operations) or connected to any of the row output buses. It is always available as an input to the next higher digital PSoC block and to the analog column clock selection multiplexers, regardless of the setting of this parameter. This parameter appears only for members of the CY8C27/24/22xxx family of PSoC devices.

The compare out is a new function for the 8-bit timer as they have listed above. It outputs a low on the reloading of the counter from the period register, then changes to a high when the compare state becomes true. You notice the statement that this output is available to the next higher digital block. One of the input options for digital blocks is to choose the output of the block immediately to its left. This option applies to the clock source of the timer. Since we placed the timer in DBB0, this option was not available among our choices for the timer source. If you were to place another timer in DBB1, then one of the clock options for that timer would be DBB0. The clock for that timer could then come from the state of compare state of this timer regardless of whether you choose none or any of the other output options. Note: Both timers would have to be started in order for the second timer to run correctly.

Period

This parameter sets the period of the timer. Allowed values are between 0 and 255. This value is loaded into the period register. The period is automatically reloaded when the counter reaches zero or the timer is enabled from the disabled state. This value may be modified using the API.

I have set the period value to a 99. The timer expires on a carry condition, not a zero condition. So by setting the period value to a 99, I have effectively divided the input clock value by 100 (period value +1).

CompareValue

This parameter sets the count point in the timer period when a compare event is triggered. This value is loaded into the compare register. Allowed values are between zero and the period value. This value may be modified using the API.

CompareType

This parameter sets the compare function type to *less than* or *less than or equal to*.

InterruptType

This parameter specifies whether the terminal count event or the compare event triggers the interrupt. The interrupt is enabled using the API.

ClockSync

In the PSoC devices, digital blocks may provide clock sources in addition to the system clocks. Digital clock sources may even be chained in ripple fashion. This introduces skew with respect to the system clocks. These skews are more critical in the CY8C27/24/22xxx PSoC device families because of various data-path optimizations, particularly those applied to the system buses. This parameter may be used to control clock skew and ensure proper operation when reading and writing PSoC block register values.

Appropriate values for this parameter should be determined from the following table:

ClockSync Value	Use
Sync to SysClk	Use this setting for any 24 MHz (SysClk)-derived clock source that is divided by two or more. Examples include VC1, VC2, VC3 (when VC3 is driven by SysClk), 32 KHz, and digital PSoC blocks with SysClk-based sources. Externally generated clock sources should also use this value to ensure that proper synchronization occurs.
Sync to SysClk*2	Use this setting for any 48 MHz (SysClk*2)-based clock unless the resulting frequency is 48 MHz (in other words, when the product of all divisors is 1).
Use SysClk Direct	Use when a 24 MHz (SysClk/1) clock is desired. This does not actually perform synchronization, but provides low-skew access to the system clock. If selected, this option overrides the setting of the Clock parameter, above. It should always be used instead of VC1, VC2, VC3 or digital blocks where the net result of all dividers in combination produces a 24 MHz output.
Unsynchronized	Use when the 48 MHz (SysClk*2) input is selected. Use when unsynchronized inputs are desired. In general, this use is advisable only when interrupt generation is the sole application of the counter.

TC_PulseWidth

This parameter provides the means of specifying whether the terminal count output pulse is one clock cycle wide or one half clock cycle wide.

InvertEnable

This parameter determines the sense of the enable input signal. When **Normal** is selected, the enable input is active-high. Selecting **Invert** causes the sense to be interpreted as active-low. InvertEnable applies only to the CY8C27/24/22xxx family of PSoC devices.

The only set up remaining for our simple project in the Device Editor Configuration view is to set up the I/O pin that will flash our LED. I have chosen P0[7] for the LED drive pin, so that pin will need to be in a state where it can sink current. Select **Strong** from the drop-down combo box titled drive on the same row as the P0[7] Port description. Make sure that the select option is set to standard CPU. I've also changed the name of this pin to LED. This doesn't affect the operation of the pin at all, but allows us to change the name of the pin as it appears on the data sheet and in the part outline section of Device Editor Configuration view.

Now that we have added all the modules that we will use in our group, it's time to start editing the necessary source code to make our project come together. Before we start editing, we need to tell PSoC Designer to generate the associated source files and code needed to setup the modules and interconnects according to our selections in the Interconnect View. Selecting **Generate Application** from the **Config** menu does this.

Application Editor

Select **Application Editor** from the **Config** menu. This will take you to a text editor in which to edit the source files of your project. On the left is a project explorer that will show you an expanding tree structure of the files in your project. These files were generated when you chose the **Generate Application** action. If you need to return to add or change anything from the Device Editor section of your design, then you should select the **Generate Application** action again so that your changes are reflected in these files. It is important also to follow the guidelines inside these source files regarding where to place your code or you will find that your hard work can disappear with the click of the mouse.

The bottom of the screen shows the output of the compiler. There are several subsections of the output area of the Application Editor. You will notice the tabs that will allow you to navigate through these views, such as, **Build, Debug, Program, Find in Files,** and so on. PsoC Designer will automatically switch to the subsection needed upon performing certain actions.

The top right and largest part of the Application Editor is for editing the project files. Double-clicking on a file in the project explorer will open that file for editing.

Project File Sections

The source files section contains the boot.asm and main.asm files. The file boot.asm is where the chip will start executing code on power-up. It will initialize the resources of the chip as you've described in the Device Editor and will then hand control over to the main.asm file. The main.asm file is your entry point for the micro. In main. asm, you can add your own code or call other files as needed to execute code.

The headers section will contain any *.inc or *.h (include, header, or library) files that you want to have associated with your files in the source files section. Typically, when you use the INCLUDE <filename> directive in your project, it will add a link to those included files here. I have seen PSoC Designer sometimes choose to put them in the external headers section also. I don't know what criteria they use to choose where to put this association, so I'm not sure why this may happen, but I have seen it happen on some projects. I believe that it sometimes feels that the files I am including aren't in the current project directory.

The library source section will contain the files associated with the modules that you have added and placed in your project. Since this project is only using the Timer8 module, this list is very small. The files timer8_1.asm and timer8_1int.asm are associated with this timer. The timer8_1.asm includes the routines associated with starting, stopping, and configuring the timer operation. The file timer8_1int.asm contains the interrupt service routine for the timer.

The other two files in the library source section are titled psocconfig.asm and psocconfigtbl.asm. These two files are referenced during the execution of boot.asm to configure the control registers of the PSoC for desired operation.

The library headers section contains include and header files that are associated with your selected modules. Under this section you will see a timer8_1.inc and a timer8_1.h file. They are generated with the intention that you will have an include file for both assembly language and a header file for writing your project in C if you so desire.

You will also notice two other pairs of files in the library headers section. The globalparams.inc and globalparams.h files contain equate statements associated with the resources of the PSoC micro in general. The psocapi.inc and psocapi.h files are generated for your convenience. They will include all of the module include files in your project, so that you won't have to include them individually. For example, let's say that your project has filled up all eight digital block with various modules. In the main.asm file you want to utilize equates made in the associated include files for each module. Rather than add an include statement for each particular modules include or header file, you can simply include psocapi.inc in the main.asm file and then reference all the module include files. This has been a great addition to the later releases of PSoC Designer.

The external headers section references an include file that doesn't reside in your project directory. You should see one file in this section, m8c.inc. It contains some macros that are utilized often enough that Cypress decided to make them available in all projects. The actual file that it's referencing resides in the PSoC Designer application directory.

The final file that you will see in your project view is the flashsecurity.txt file. This file will allow you to set the security of each block of Flash on your PSoC micro. The default setting is a protected state. You will need to change this if you plan on being able to read the Flash from an external programmer or plan to program Flash without doing a mass erase first. If you are using the EEPROM module, you will need to unprotect the associated block of memory for the module to function properly. The default setting for all blocks of memory is W or 'Full' security.

Some Important Files

boot.asm

This file is regenerated every time you use the **Generate Application** action. Any changes that you have made to this file will be lost when the file is regenerated. If a change must be made to the boot area of your project, you should make the change to the boot.tpl file in your projects directory. It will alter how boot.asm is regenerated every time you use the **Generate Application** function. Change this file with extreme care. It's always good practice to make a backup copy of the project you are working on to make sure that you can restore it back to a working condition if needed.

Let's double-click on boot.asm from the source files section of the project explorer to open it. Now press **Ctrl** + **G** on your keyboard, type in 100, and then click on OK to go to line 100 of this file. Somewhere close to line 100, you should see an org 20h statement. You will notice from the comment that it contains the interrupt vector for the DBB00 PSoC block. This block is where we placed our 8-bit timer. There should be an ljmp instruction at this line followed by the PSoC Designer-generated location of the interrupt service routine. We will modify the code in that interrupt service routine to set up the base timing structure of our project. There is a possibility that the aforementioned instruction is really at a different line in the boot.asm file. Don't panic. PSoC Designer is continually working to improve the operation of code generation and may have made some changes to how the boot.asm file is structured. If you are having trouble finding the location of this section, use the **Ctrl** + **F** shortcut to bring up a **Find** window. Then search for the instances of the 20h within the file and you should find it without any trouble.

You will notice around this jump statement that there are declarations for the various PSoC blocks and other associated interrupts within the project. Some interrupt vectors may need to be set up manually and that can be done in this file by placing an ljmp instruction similar to the one we just examined after the appropriate org statement. Remember however, that this file is regenerated every time you make a change in the Device Editor section of PSoC Designer. That will regenerate this file and will overwrite what you have done. You can either be patient and keep making the necessary changes after regeneration, or you can make the necessary changes in the boot.tpl file located in your project directory. If you choose to make the changes to the boot.tpl file, then make sure to make a backup of your project and make sure to examine the boot.asm file after regeneration to make sure that the desired changes have been made to the boot.asm file appropriately.

psocconfigtbl.asm

The psocconfigtbl.asm file is found in the library source section of the project explorer. Open this file and examine its contents. This file has the control register settings for your project. There are two different methods employed by PSoC Designer to load the control registers. There is a direct loading of the registers using the mov instruction. There is also the method of indexing into a table. PSoC Designer might use the former to speed up loading of the registers or the latter to conserve space in Flash memory. There are some options to determine which type is used in the **Project → Settings → Device Editor** option from the **Project** menu. You will see this option in the Configuration Initialization section of that window.

This is a very important file to know your way around. You can use it as a great teaching tool when you want to start changing the values of the control registers without using the API calls. An effective knowledge of handling the control registers will help you turn the PSoC parts into very powerful tools. If that seems a little unlikely, you might find that you need to have a working knowledge of setting up these registers just to make your project accomplish its minimum function.

Each line has a comment describing what register is being loaded. Whether using the table method or a direct loading, the first hex number is the location of the register and the second number is the value being loaded into that register. Pay particular attention to the registers associated with your current project. In this particular project, we placed the 8-bit timer in the DBB00 block. You will notice the registers associated with this block near the end of the psocconfigtbl.asm file. The name (Timer8_1) that we gave the block is included in the comment for your convenience. If you are having trouble finding the section, use the **Ctrl + F** find window to search for DBB00.

Variable Declaration

Now that we have the project started, let's get to work putting the code into place. For a tidy organization, I prefer to put the variable declarations used in this project in their own file. In order to add a new file to your project, select **New File** from the

File menu. This opens a dialog box that allows you to create a file of a particular type and will add that file to the project for you. Note that if you want to add a file that already exists, then you choose **Add to Project → Files** from the **Project** menu.

We are going to add an M8C Assembler Source file, so click that selection and name the file ram.asm. After clicking OK, you will notice that PSoC Designer will create this file and open it for editing immediately. PSoC Designer has also added the project to the group source files in the project explorer if you left the default option to add the file to the project checked.

Inside of ram.asm, we are going to define a few variables for this project using the blk directive. Before doing this, however, we will need to define the area inside of the ram.asm file as RAM. We will do this using the area directive.

The area directive will allow you to define a section as being part of RAM or ROM. It also allows you to define the area in an absolute location, or define it as a relocatable section. Further information about the area directive can be found in the Assembler Directives section of the PSoC Designer help file. We can utilize the defaults of the area directive and add a simple declaration at the top of the ram.asm file.

Area bss(RAM)

This instance of the area directive tells the compiler that we are entering a RAM section that has the name bss. The reason that I chose the name bss is because it is the name of the RAM section that is declared in boot.asm. This section will contain most RAM that is reserved by your blocks for their proper operation. Since I'm using the default options of the area directive, the RAM declarations that we make using the block directive will be appended into the bss area of RAM. The bss area of RAM will start at address 0 and will grow upward. Remember that the stack of the PSoC grows towards higher addresses, so keeping the RAM closely grouped together towards the lowest addresses possible allows us to leave the most room possible for the stack to grow. PSoC Designer will assign the stack pointer to start at the location following the bss area in our project.

Now that the RAM section has been declared, it's time to declare some variables using the blk directive. The blk directive reserves a block of RAM. The size of the block is determined by the argument following the directive. In this project, all variables will only be 1 byte. I've listed the variables we need below, including the area directive.

```
Area   bss(RAM)

timeflags::        blk   1   ;register divided into various flags
tsecond_count::    blk   1   ;counter for a tenth of a second
second_count::     blk   1   ;counter for a second
```

You will notice that after each variable name, I have put two colons. This is a short way of declaring that name as global. There is another way of doing this: using the EXPORT directive. This may be preferred in some cases. For example, the EXPORT directive is used in the library files. It's nice to be able to see all available routines simply by examining the top of the library file, however, in this case I prefer to use the double colon method since everything within this file will be used globally.

Constant Declarations

In order to use our timeflags register effectively, we need to set up some constant declarations. Since these constant values will be used in multiple files, I prefer to include a file that contains all the global constant declarations. Create a file using the same method used for creating the ram.asm file, but use the M8C Assembler Include File as the file type. Let's name this file constants.inc.

When you press OK to create this file, you will notice that the file is not immediately added to the project explorer. This is because none of the files in the project are currently referencing this file. You can change this by adding an include statement to the top of the main.asm. Put the include statement under the include "PSoCAPI.inc" statement. Your new line will read as follows:

```
include "constants.inc"
```

Once you have added this line, press F7 to build the project. You can also select **Build** from the **Build** menu. In the output section of Application Editor, you will see the results of the build. If you have done everything right so far, you should have zero errors and zero warnings. After building the project for the first time, you will see a + sign show up on the headers group of the project explorer. Clicking on this + sign reveals that the constants.inc file has now been added, since it is referenced in main.asm.

Inside the constants.inc file, we are going to define the timing bits that will be used to set up the frequency needed for our project. So let's add the following lines to constants.inc:

```
      include  "PSoCAPI.inc"
   ;Port Bit definitions
   LEDport:        equ   PRT0DR        ;Port register used for LED
   LED:            equ   128           ;Port mask used for LED

   ;timeflags bits
   tsecondflag:    equ   1             ;mask for tenth second occurrence
   secondflag:     equ   2             ;mask for second occurrence
```

There isn't a need for the double colon in this file, since I will be using the INCLUDE directive to include this file within my other project files. You will notice that I'm including the PSoCAPI.inc file within this file. This will let me reference other constants that apply to my library files within this file. It will also allow me to only have one include statement for any other source files that I add to my project.

Timer Interrupt

Now it's time to set up some timing. Open the timer8_1int.asm file from the library source group of Project Explorer.

At the top of the file, you will see some include statements that reference include files that already exist within the project. Resist the temptation to add your statement to these for the reason I will give you later on.

Underneath these include statements, you will see export statements of all the routines that are available within this file. There is only one in this file, but if you open the timer8_1.asm file, you will notice several routines. You may notice that there are two separate labels for each subroutine in the timer8_1.asm file. That is because one label is used if you are calling the routine in assembly. The other label is used if you are calling the name in C.

In the timer8_1int.asm file, there is only one label for the interrupt routine as this routine isn't typically called from your files, but is used as a vector label for the interrupt routine in boot.asm.

Near the top of this file, you will see the following lines:

```
;----------------------------------------------------
; Insert your custom declarations below this banner
;----------------------------------------------------
```

Just underneath these lines add the following statement:

```
include     "constants.inc"
```

This makes all the declarations of the constants.inc file available for use in this file. It's important for you to place this include statement in the section allowed for your custom declarations. If you place it outside of this area and not in another reserved area, then it will be overwritten the next time you use the generate application information and you will be going to backup files to retrieve your code.

Continue on further down the file until you get to the interrupt label _Timer8_1_ISR.

Under this label you will see the following comments:

```
;@PSoC_UserCode_BODY@ (Do not change this line.)
   ;----------------------------------------------------
   ; Insert your custom code below this banner
   ;----------------------------------------------------
   ;    NOTE: interrupt service routines must preserve
   ;    the values of the A and X CPU registers.

   ;----------------------------------------------------
   ; Insert your custom code above this banner
   ;----------------------------------------------------
;@PSoC_UserCode_END@ (Do not change this line.)
```

This is another section that is set aside not to be overwritten with the generate application function. We'll add our code just above the **Insert your custom code above this banner** section. After we are done, it should look like the following:

```
;@PSoC_UserCode_BODY@ (Do not change this line.)
   ;----------------------------------------------------
   ; Insert your custom code below this banner
   ;----------------------------------------------------
   ;    NOTE: interrupt service routines must preserve
   ;    the values of the A and X CPU registers.
   push  a
```

```
        push    x                               ;preserve a and x resgisters
        inc         [tsecond_count]             ;increment tenth second count
        cmp         [tsecond_count],100         ;until it reaches 100
        jc          EndInt
        mov         [tsecond_count],0           ;reset tenth second count
        or          [timeflags],tsecondflag     ;set tenth second flag

        inc         [second_count]              ;increment second count
        cmp         [second_count],10           ;until it reaches 10
        jc          EndInt
        mov         [second_count],0            ;reset second count
        or          [timeflags],secondflag      ;set second flag
EndInt:
        pop     x                               ;restore a and x registers
        pop     a


        ;-------------------------------------------------
        ; Insert your custom code above this banner
        ;-------------------------------------------------
        ;@PSoC_UserCode_END@ (Do not change this line.)
```

First, you will notice that I follow the advice of PSoC Designer and preserve the A and X registers. I am aware that these registers aren't used in the routine, but I will do it anyway out of good habit since I may want to expand this routine later one to use those registers and don't want to be caught forgetting to add the context saving. The saving of the flags is done automatically with the interrupt.

The rest of the interrupt is pretty self-explanatory. This timer was set up to interrupt every 1.0 msec. The object is to count how many times the interrupt has occurred. After 100 occurrences of the interrupt I will set the tsecondflag to indicate that one-tenth of a second has expired. After 10 occurrences of the tenth-second section, I will set the secondflag to show that one second has expired.

I'm not going to add all the code that I would like execute every tenth of a second or every second within the interrupt. I do that for a couple of reasons. First of all, I'm within an interrupt. I don't want to bog down the interrupt with a bunch of code that causes me to add a lot of context saving. Also, I don't want to spend a lot of execution time within my interrupt. I want to get in and get out quickly so the processor can decide what code is important enough to work on at the moment. Since the interrupt has set flags showing what time has expired, I can use the flags back in the main.asm file to determine when I want to perform certain tasks.

Main.asm

When the PSoC powers up, it looks at the vector located at address 0000h. It then starts executing code at the location jumped to by that vector. These two locations are declared in boot.asm. If you look towards the top of boot.asm just under the equate statements and export statements, you will find an org statement for the reset vector. It's at line 69 in my project, but may be at a different location in your project.

Under that org statement, you will see a ljmp instruction that takes processor operation to the __Start label. This location is just under the vector table at location 68h. Operation continues at this point to load the default table values (as defined by our device editor choices) in the appropriate control registers. Near the bottom of the boot.asm file, you file a lcall _main instruction that takes us into main.asm.

You will notice that control transfer to the _main location (located in main.asm) is done with the lcall statement rather than a ljmp. This allows a safety net so that if you hit an unexpected return without an additional call in main.asm, you will simply return to boot.asm and stay in an endless loop. This would prevent a reset in this condition. If you prefer for the project to reset in this kind of condition, then you will need to modify boot.asm. Remember that a change to the boot.tpl template file in your project directory allows you to change the way boot.asm is generated and such changes would then be preserved the next time that you use the **Generate Application** function.

In the main.asm file, you will see that the _main label is already in the file along with a return statement underneath it that will return code operation to that endless loop within boot.asm. Our mainloop section will lie in between the _main label and the return statement. I've included the entire mainloop for your reference.

```
call  Timer8_1_EnableInt    ;Enable the Timer interrupt
M8C_EnableGInt              ;Enable interrupts
call  Timer8_1_Start        ;Start the timer
```

```
     ;Add any other initialization here
mainloop:

     tst    [timeflags],tsecondflag  ;Test the timer interrupt flag
     jz     mainloop                 ;Loop until 0.1 seconds has expired
     and    [timeflags],~tsecondflag ;clear the timer flag
     ;This area executes every 0.1 seconds

     tst    [timeflags],secondflag   ;Test the timer interrupt flag
     jz     mainloop                 ;Loop until 1.0 seconds has expired
     and    [timeflags],~secondflag  ;clear the timer flag
     ;This area executes every 1.0 seconds
     xor    reg[LEDport],LED         ;Toggle the LED on an off

     jmp    mainloop
```

Before I drop down into mainloop, I need to make sure the 8-bit timer is running and its interrupt is able to happen so that the timer's interrupt service routine can set the appropriate timing flags. I accomplish this by calling PSoC Designer-generated API functions that reside in the timer8_1.asm file in the lib directory. These subroutines are declared as global with the export directive in their respective file, so main.asm can find their location to complete the call statements. There is no need to define these routines as externally defined with the main.asm file. This might vary a little from the assemblers that you have used before but PSoC Designer automatically will assume that you are calling an externally defined location and will generate an error if it doesn't find this location declared in another file included in the project.

You should add any further initialization that is needed for the project, before dropping in to the main loop. This might include a section that initializes RAM. (Note that adding C language support to a project will include a section of boot.asm that will clear all registers that belong to the bss segment.)

Inside the mainloop, I've used the tst instruction to determine if the timer has expired enough times to set the tsecondflag bit indicating that one tenth of a second has already expired. If this flag has not yet been set then the processor jumps back to the mainloop label. Once the tenth second has expired and the interrupt service routine has set the tsecondflag bit, then the processor continues execution after the jz statement, clears the tsecondflag bit and executes code in the area defined as happening every tenth of a second.

Immediately below this area, I've set up a similar test and conditional jump statement to organize an area that executes every one second. Once that area has been processed, the PSoC jumps back to the main loop label. You will notice that the mainloop area is structured such that all happenings that occur on the one-second time scale are executed following the tenth-second tasks that occur at that same time. This allows an order of operations that might be useful in your design.

In the one-second area, I've added a single line statement to exclusive-OR the port 0 data register with the flag as defined in my constants.inc file. This will toggle the output pin that is tied to the LED and will give me my 0.5 Hz blink of the LED. I could easily change the frequency to 5 Hz by moving this statement into the tenth-second area of the main loop.

Building the Project

If you press F7 on your keyboard, the project will be assembled and linked and ready for you to program. If you watch the output section of Application Editor, it will report the happenings of the build process. You will also see an indication of how full the device is. This is for reference only. It's not perfect, but gives you an idea of where your code usage stands. You should also see any errors that have cropped up during the build process. If errors exist, please check the file with errors for any potential problems. If your project had an error at this point, my guess would be that you spelled something wrong. Note that any errors in your project will prevent the generation of the output file(s).

Now you are ready to program your chip and put it into your project. If everything is working, then your LED should be blinking at a slow rate of 0.5 Hz.

Goals of This Exercise

I know that this project could be accomplished without writing any additional code. I could have used a larger timer block or utilized VC3 to allow me to blink the LED directly by connecting the LED to the output of a timer that is running sufficiently slowly to meet the design requirements.

I decided to present the project in the way that I have so as to offer a good base for building larger projects. The areas of code that are executed every tenth second or every one second allow you a standard template for building a good visual interface, without the need of having several interrupts enabled. In my design experience, most people consider anything that occurs in the first 0.1 seconds as being fairly instantaneous, hence the tenth-second area. The one-second area gives you a location to place code that needs to keep plodding along at a regular rate, but isn't essential to have updated faster than once per second.

The project above was also designed to work on every PSoC part that is currently shipping. It is possible that future PSoC devices might require modifications from the design that I've described.

Limited Analog System

Cypress has produced some new analog parts that have limited analog capability. These parts are targeting simpler, low cost systems where the full extent of the analog capability is not needed. Examples of this limited analog system are found in the 21xxx parts. These parts have four digital blocks and four analog type E blocks. These analog blocks are arranged in two columns. Each column is comprised of a continuous time block and a switched capacitor block. This arrangement is shown in Figure C-1.

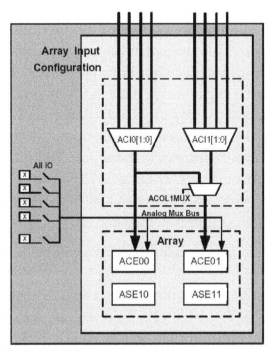

Figure C-1: Analog Type E Block Arrangement

ACE00 and ACE01 are the continuous time analog blocks. ASE10 and ASE11 are the switched capacitor analog blocks. These blocks provide you with two 8-bit analog-to-digital converters and two-input comparators which compare two different analog I/O pins or a single-input comparator that compares a signal with a digital-to-analog reference level.

There are some important limitations to note with these continuous analog blocks compared to the analog blocks present on other PSoC devices. The architecture of the type E continuous time analog block is shown here in Figure C-2.

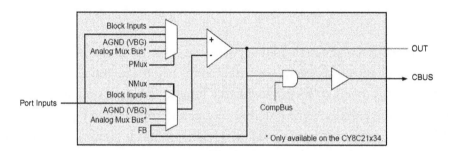

Figure C-2: Type E Continuous Time Analog Block

There are no feedback resistor blocks for the ACExx blocks like you find on the ACAxx andACBxx blocks. This means that you can not set a variable gain on the device. The ACExx analog blocks offer the unity gain of one, which is a direct connection between the output and inverting input of the internal operational amplifier. The ACExx analog blocks also offer no feedback connection between the output and the inverting input that then makes the operational amplifier function as a comparator. The separate low-power comparator that was introduced with the newer 27xxx and similar parts is not present in the 21xxx series. The ACExx blocks were designed to be low power in their operation however and the power that is consumed by the ACExx may suffice for your project requirements.

Figure C-2 shows the various inputs available to the continuous time block. The port inputs shown are structured like the analog multiplexers on the other PSoC microcontrollers. It even has a die temperature input for ACE01 although it isn't shown in the visual configuration. You will notice, however, that the type E blocks offer a new input called the *analog mux bus*. The analog mux bus is also seen in Figure C-1.

This analog mux bus has the potential to connect to any one of the I/O pins, thereby allowing you to input an analog signal in from any of the I/O pins.

The ASExx switched cap blocks are also different from the ASAxx, ASBxx, ASCxx, and ASDxx switched cap blocks. They only have one configuration, which consists of a constant current source and an integrator which forms a DAC. This DAC is fed back to the continuous analog block to form an analog-to-digital converter. The only configurations under your control are inputs, outputs, and the speed of the integrator. An illustration of the type E switched cap block is given in Figure C-3.

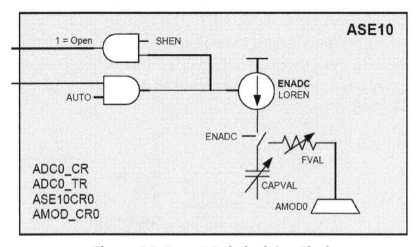

Figure C-3: Type E Switched Cap Block

The analog mux bus does allow a connection to any I/O pin. Connection to the analog mux bus is done by setting the appropriate bit in the analog mux port bit enables registers. There are four of these registers. They are named MUX_CR0, MUX_CR1, MUX_CR2 and MUX_CR3. Each register corresponds to the port with the same designation. Setting the appropriate bit in one of these registers will connect the I/O pin to the analog mux bus. There is no limitation on how many I/O pins are connected at any one time. However, you should take note that when you connect multiple I/O pins to the analog mux bus you are effectively tying those pins together, so that the signals of all the I/O pins are seen on the bus. This may not be desirable if you are trying to isolate signals from each pin individually. If this is what you are trying to do then you need to make sure that only one pin is connected to the analog mux bus at a time. You should also note that the analog mux bus circuitry is implemented in parallel with the normal GPIO circuitry. If you don't have the pin set in the analog input state, then the other effects of the GPIO settings will be reflected on the I/O bus, which could cause contention between signals.

The analog bus also has specific charging capabilities that are used in capacitive sensing. There is some break-before-make circuitry (Figure C-4) that is connected to each I/O pin. This circuitry is controlled via the AMUXCFG register. The default state of this register always leaves the I/O pin connected to the analog mux bus if the MUX_CRx.n register for that pin is asserted as described previously. However, by enabling the precharge clock, you will set up the circuitry to disable the connection from the analog mux bus while the pin is charging. The pin will be reconnected to the analog mux bus after charging is complete. The idea is that you have a capacitor acting as an integrator on P0[3] or P0[1]. These two pins have the option of remaining connected to the analog mux bus at all times via bits 5 and 4 in the AMUXCFG register.

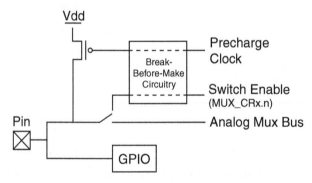

Figure C-4: Break-Before-Make Circuitry

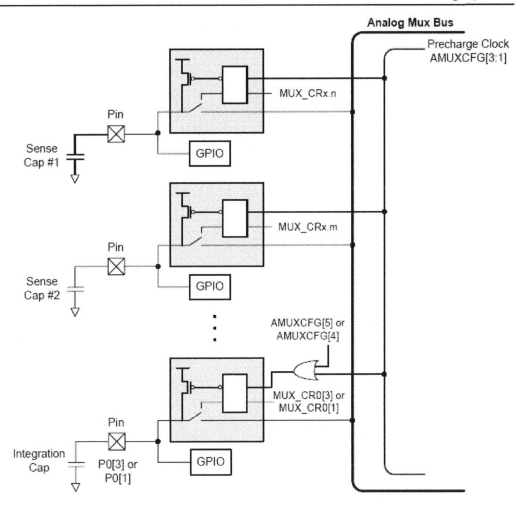

Figure C-5: Capacitive Sensing

The idea is to discharge the integration capacitor via the I/O pin circuitry, in other words, make it a strong drive low signal. Then you turn that I/O pin back into an analog input and start one of the capacitors charging. You measure how many cycles (as a length of time) it takes for the sense cap to charge the integration cap past a known point. That is used as a measurement of the capacitance of the sense pin. More information in available in the application notes section from Cypress.

About the Author

Robert Ashby designs electronics for the fitness industry and a variety of consulting projects. He received a BS in electronics and computer science from Utah State University. Robert has taught courses at Utah State University and Bridgerland Applied Technology Center. His affinity for his farm boy upbringing and wide experience with electronics has provided him with endless opportunities for uniting embedded design with mechanical systems. He lives in Beaver Dam, Utah with his wife Camille and three children.

Index

Printed and bound by CPI Group (UK) Ltd, Croydon, CR0 4YY

03/10/2024

01040336-0002